STRENGTHENING NATIONAL PUBLIC HEALTH PREPAREDNESS AND RESPONSE TO CHEMICAL, BIOLOGICAL AND RADIOLOGICAL AGENT THREATS

NATO Security through Science Series

This Series presents the results of scientific meetings supported under the NATO Programme for Security through Science (STS).

Meetings supported by the NATO STS Programme are in security-related priority areas of Defence Against Terrorism or Countering Other Threats to Security. The types of meeting supported are generally "Advanced Study Institutes" and "Advanced Research Workshops". The NATO STS Series collects together the results of these meetings. The meetings are co-organized by scientists from NATO countries and scientists from NATO's "Partner" or "Mediterranean Dialogue" countries. The observations and recommendations made at the meetings, as well as the contents of the volumes in the Series, reflect those of participants and contributors only; they should not necessarily be regarded as reflecting NATO views or policy.

Advanced Study Institutes (ASI) are high-level tutorial courses to convey the latest developments in a subject to an advanced-level audience.

Advanced Research Workshops (ARW) are expert meetings where an intense but informal exchange of views at the frontiers of a subject aims at identifying directions for future action.

Following a transformation of the programme in 2004 the Series has been re-named and re-organised. Recent volumes on topics not related to security, which result from meetings supported under the programme earlier, may be found in the NATO Science Series.

The Series is published by IOS Press, Amsterdam, and Springer Science and Business Media, Dordrecht, in conjunction with the NATO Public Diplomacy Division.

Sub-Series

A.	Chemistry and Biology	Springer Science and Business Media
B.	Physics and Biophysics	Springer Science and Business Media
C.	Environmental Security	Springer Science and Business Media
D.	Information and Communication Security	IOS Press
E.	Human and Societal Dynamics	IOS Press

http://www.nato.int/science
http://www.springer.com
http://www.iospress.nl

Sub-Series E: Human and Societal Dynamics – Vol. 20 ISSN: 1574-5597

Strengthening National Public Health Preparedness and Response to Chemical, Biological and Radiological Agent Threats

Edited by

C.E. Cummings

Drexel University School of Public Health, Philadelphia, PA, USA

and

E. Stikova

University "St Cyril and Methodius", Faculty of Medicine, Skopje, Republic of Macedonia

IOS Press

Amsterdam • Berlin • Oxford • Tokyo • Washington, DC

Published in cooperation with NATO Public Diplomacy Division

Proceedings of the NATO Advanced Study Institute on Strengthening National Public Health Preparedness and Response for Chemical, Biological and Radiological Agents Threats
Skopje, Republic of Macedonia
19–29 June 2006

ISBN 978-1-58603-744-4
Library of Congress Control Number: 2007930109

Publisher
IOS Press
Nieuwe Hemweg 6B
1013 BG Amsterdam
Netherlands
fax: +31 20 687 0019
e-mail: order@iospress.nl

Distributor in the UK and Ireland
Gazelle Books Services Ltd.
White Cross Mills
Hightown
Lancaster LA1 4XS
United Kingdom
fax: +44 1524 63232
e-mail: sales@gazellebooks.co.uk

Distributor in the USA and Canada
IOS Press, Inc.
4502 Rachael Manor Drive
Fairfax, VA 22032
USA
fax: +1 703 323 3668
e-mail: iosbooks@iospress.com

PRINTED IN THE NETHERLANDS

Strengthening National Public Health Preparedness and Response to Chemical, Biological and Radiological Agent Threats. Edited by C.E. Cummings and E. Stikova.
IOS Press, 2007.

Preface

We are glad to present the proceedings of the North Atlantic Treaty Organization's Advanced Study Institute, *Strengthening National Public Health Preparedness and Response for Chemical, Biological and Radiological Agents Threats*, held in Skopje, Republic of Macedonia, on June 19 to 29, 2006. We are grateful to the NATO Science Programme both for sponsoring its ASI series, and for recognizing the global importance of public health preparedness against terrorism and weapons of mass destruction.

With gratitude we recognize and thank those organizations whose funding made the Institute possible. Funding for participants came from NATO in the form of a conference grant. The Institute was sponsored by the University St. Cyril and Methodius, Skopje, Macedonia, which also provided logistical support, a hospitable atmosphere, tours of Macedonian national health facilities, and host country events. The Hotel Continental, Skopje, was our venue; it provided lodging, meals, all meeting rooms, and a fine overall setting for the ASI. Additionally, many participants' travel costs were funded in part or in whole by their own organizations.

We were honored to have had such a fine roster of military and civilian public health experts from NATO nations and NATO Partner nations. The number of participants included a rich variety of experts in all aspects of the program, who excellently represented a broad array of eligible countries. A core panel of participants contributed to final discussions and conclusions of the Institute. New relationships and collaborations were forged during this Institute. The recorded product is represented in this book, but perhaps more valuable were the new relationships and potential joint projects made possible in Skopje.

The scheduled format of the ASI was as follows: there were two morning sessions and two afternoon sessions daily. Presentations were organized according to scientific discipline, with these groupings – the American Medical Association's (AMA) two-day Basic Disaster Life Support course; biological, radiological and chemical threats; computer simulation and emergency planning; information security – approach and management; biodefense; risk communication; health policy; public health law and preparedness; "special needs" populations in preparedness planning; special topics; and psychological aspects of disasters. There was also a brief avian influenza table-top exercise.

We believe that this report's data, recommendations, and conclusions are neither final nor all-inclusive. The participants' group experience and knowledge regarding public health emergencies has continued, driven by those ongoing, always-changing terror threats that promise to kill, maim and disrupt the lives of the civilized world. We plan continued dialogue on the recommendations found herein, and hope that future similar workshops will occur.

Finally, we urge that that this Institute's recommendations are accepted and implemented by clinicians, researchers, and other scientists motivated by special interest in public health preparedness, as well as by national and NATO leaders and policy makers who are positioned to make a difference. Public health response to emergencies requires extensive, coordinated, considered efforts of the combined military and civilian public health resources of all NATO nations and NATO Partner nations.

Curtis Cummings
Carol Larach

Acknowledgements

We would like to take this opportunity to thank all of those who assisted in the editing of these proceedings, in particular, Elin Gursky, who provided valuable expertise in this process. We acknowledge also the editorial board for all their work reviewing and editing and providing critical feedback:

Carol Larach, Managing Editor

Marija Caplinskiene
Elin Gursky
Faina Linkov
Alessandra Rossodivita
Eugene Shubnikov
Andrei Trufanov
Conrad Volz

Curtis E. Cummings
Elisaveta Stikova

List of Lecturers and Students

Akbas, Eten
Mersin University, Department of Medical Biology and Genetics, Yenisehir Campus, Mersin, Turkey

Alvelo, Jose
Northrop Grumman Corporation, Alexandria, Virginia, USA

Bakanidze, Lela
National Center for Disease Control and Medical Statistics of Georgia, 9. M. Astiani, St. Tbilisi, 0177, Georgia

Berjohn, Catherine
Drexel University School of Public Health, 1505 Race Street, 11th Floor, Philadelphia, PA, USA 19102

Bosevska, Golubinka
National Public Health Institute, U1. 50 Divizija bb. 1000, Skopje, Republic of Macedonia

Buchavyy, Yuriy
National Mining University of Ukraine, Y. Savchenko Str. 6B/16, Dnepzopetrovsk 49006, Ukraine

Caplinskiene, Marija
Mykolas Romeris University, Institute of Forensic Medicine, Zukausko 12, LT-08234, Vilnius, Lithuania

Chukaliev, Ordan
National Public Health Institute, Faculty of Agricultrure Science and Food, Bul. Aleksandar Makedonski bb. 1000, Skopje, Republic of Macedonia

Cohen, Bruce
Naval Operational Medicine Institute, Pensacola, Florida, USA

Coule, Phillip
AMA – Medical College of Georgia, 110 15th Street, AF-20259, Ausgusta, Georgia, USA

Cummings, Curtis
Drexel University School of Public Health, 1505 Race Street, 13th Floor, Philadelphia, PA, USA

Dimovska, Mirjana
National Public Health Institute, U1. 50 Divizija bb. 1000, Skopje, Republic of Macedonia

Gudzenko, Nataliya
Research Center for Radiation Medicine, Melnikova 53, Kiev 04050, Ukraine

Guidotti, Matteo
Institute of Molecular Sciences and Technology, CNR-ISTM, Milan, Italy

Gursky, Elin
National Strategies Support Directorate/ANSER, 2900 South Quincy Street, Suite 800, Arlington, VA, USA

Hill, Jeffrey
CMR 402, Box 1356, APO, AE

Horochena, Krista
University of Maryland, Center for Health and Homeland Security, 500 West. Baltimore Street, Baltimore, MD, USA

Horosko, Steve
FORSCOM, Fort McPherson, Gergia, USA

Kakaraskoska, Biljana
National Public Health Institute, U1. 50 Divizija bb. 1000, Skopje, Republic of Macedonia

Karadzovski, Zerako
Institute of Occupational Health, II Makedonska Brigada 43, Skopje, Macedonia 1000

Kemdrovski, Vladimir
University of St. Cyril and Methodius, National Public Health Institute U1. 50 Divizija bb. 1000 Skopje, Republic of Macedonia

Klaiman, Tamar
Temple University, 1700 North Broad Street, Suite 304, Phialdelphia, PA, USA

Kocubovski, Mihail
University of St. Cyril and Methodius, National Public Health Institute U1. 50 Divizija bb. 1000 Skopje, Republic of Macedonia

Kohal, Pamela
Drexel University School of Public Health, 1505 Race Street, 11th Floor, Philadelphia, PA, 19102, USA

Kuchma, Irina
Kharkov Medical Postgraduate Academy, Pushkinskaya, 14

Larach, Carol
Drexel University School of Public Health, 1505 Race Street, 11th Floor, Philadelphia, PA, 19102, USA

Linkov, Faina
University of Pittsburgh Cancer Institute/School of Medicine, 3512 5th Avenue, Room 312, Pittsburgh, PA, USA

Michalski, Aleksander
Military Institute of Hygiene & Epidemiology UI Lubelska 2, 24-100 Pulawy, Poland

Mijakoski, Dragan
Institute of Occupational Health, II Makedonska Brigada 43, Skopje, Macedonia 1000

Milevska-Kostova, Neda
Center for Regional Policy Research and Cooperation, Nikola Parapunov bb, Kompleks Makoteks 1 katt, POB 484, Skopje, Republic of Macedonia

Mirkova, Ekaterina
National Center for Public Health Protection, Sofia 1431, Bulgaria

Moten, Asad Ali
Drexel University School of Public Health, 1505 Race Street, 11th Floor, Philadelphia, PA, 19102, USA

Nagrebetsky, Alexander
Kiev, Ukraine

Nedelkovski, Dusan
National Public Health Institute, U1. 50 Divizija bb. 1000, Skopje, Republic of Macedonia

Niemcewicz, Marcin
Military Institute of Hygiene and Epidemiology/Biological Threats Identification and Countermeasure Center, Lubelska 2, 24-100 Pulawy, Poland

Noda, Fatima
Drexel University School of Public Health, 1505 Race Street, 11th Floor, Philadelphia, PA, 19102, USA

Ranghieri, Massimo
SMOM Auxiliary Corps of the Italian Army, 1° Reparto, Milan, Italy

Rizvi, Ali
Drexel University School of Public Health, 1505 Race Street, 11th Floor, Philadelphia, PA 19102, USA

Robinson, Donald
Department of Systems and Information Engineering University of Virginia, Charlottesville, VA, USA

Rossodivita, Alessandra
Department of Cardiothoracic and Vascular Diseases, San Raffaele Hospital, Milan, Italy

Rumm, Peter
Center for Devices and Radiological Health, Food and Drug Administration, US Department of Health and Services, 9600 Corporate Drive, Rockville, MD, 20850-1000, USA

Shubnikov, Evgeny
Institute of Internal Medicine, P.O. Box 668, Novosibirsk 630090, Russian Federation

Smith, Kim
CDC/NCEH/DLS – Organic Analytical Toxicological, 4770 Budford Highway, Atlanta, Georgia,USA

Spiroski, Igor
National Public Health Institute, U1. 50 Divizija bb. 1000, Skopje, Republic of Macedonia

Stafilov, Trajce
University of St. Cyril and Methodius, Institute of Chemistry, Arhimedova 5, Skopje, Republic of Macedonia

Stikova, Elisaveta
University of St. Cyril and Methodius, National Public Health Institute U1. 50 Divizija bb. 1000 Skopje, Republic of Macedonia

Stoleski, Saso
Institute of Occupational Health, II Makedonska Brigada 43, Skopje, Macedonia 1000

Subbarao, Italo
American Medical Association/The Johns Hopkins University, 1900 Thames Street, Baltimore, MD, USA

Swienton, Raymond
American Medical Assocaition/The Univeristy of Texas Southwestern Medical Center, 5323 Harry Hines Blvd. Dallas, Texas, USA

Tait, Ashley
Drexel University School of Public Health, 1505 Race Street, 11th Floor, Philadelphia, PA 19102, USA

Taseva, Slagna
Police Academy, P.O. Box 103, 100 Skopje, Republic of Macedonia

Tavares, Afonso
Universidade Lusiada de Lisboa, P.O. Box 21443, 113-001 Lisboa, Portugal

Tomljanovic, Charles
Concurrent Technologies Corporation, 100 CTC Drive, Johnstown, PA 15904, USA

Trufanov, Andrey
Irkutsk State Technical University, Lermontova 83, Irkutsk, 664074, Russian Federation

Tseytlin, Eugene
Department of Biomedical Informatics, University of Pittsburgh School of Medicine, Pittsburgh, PA, USA

Varadarajan, Kartik
Drexel University School of Public Health, 1505 Race Street, 11th Floor, Philadelphia, PA 19102, USA

Volyanskiy, Andriy
Ukranian Academy of Medical Sciences, Mechnicov Science and Research Institute of Microbiology and Immunology, Ukriane, Pushkinskaya 14

Volz, Conrad
University of Pittsburgh, Department of Environmental and Occupational Health (EOH), Graduate School of Public Health, A712 Crabtree Hall (PUBHT), Pittsburgh, PA 15261

Vynograd, Nataliya
Lviv National Medical University, Pekarska Str. 69, dviv, Ukraine

Wahi, Guarav
Drexel University School of Public Health, 1505 Race Street, 11th Floor, Philadelphia, PA 19102, USA

Contents

5. Modeling and Simulation

6. Biological Agents

7. Chemical Agents

8. Radiological Agents

9. Special Topics

1. Introduction

Strengthening National Public Health Preparedness and Response to Chemical, Biological and Radiological Agent Threats. Edited by C.E. Cummings and E. Stikova.
IOS Press, 2007.

The Need for Strengthening National Public Health Preparedness and Response to CBR Agent Threats

Curtis E. CUMMINGS [a], Elisaveta STIKOVA [b], Carol S. LARACH [a] and Peter D. RUMM [c]
[a] *Drexel University School of Public Health, MS 660, Philadelphia, PA 19102-1192, USA*
[b] *University of St. Cyril and Methodius, U1. 50 Divizija bb. 1000 Skopje, Republic of Macedonia*
[c] *Center for Devices and Radiological Health, Food and Drug Administration, US Department of Health and Services, 9600 Corporate Drive, Rockville, MD, 20850-1000, USA*

Ensuring a coordinated public health, laboratory, and medical response to a natural disaster, an accidental release or a deliberate use of a chemical, biological, and radiological agent (CBR) is a high priority for developed and developing countries. To this end, a North Atlantic Treaty Organization (NATO) Advanced Study Institute (ASI), *Strengthening National Public Health Preparedness and Response to Chemical, Biological and Radiological Agent Threats*, was conducted in Skopje, Republic of Macedonia, on June 19–29, 2006. The primary aim of this ASI was to provide NATO countries and Eligible Partner countries with technical assistance on ways to enhance their national preparedness and response plans to CBR threats.

The ASI included the AMA (American Medical Association) Basic Disaster Life Support Course (BDLS) [1]; topic sections that included: preparedness systems planning, the national public health preparedness systems of countries participating in the conference (Macedonia, Bulgaria, Georgia, Italy, Lithuania, Poland, Ukraine, and the U.S.), the role of information technology and information security including the role of the University of Pittsburgh-based Supercourse Global Health Network library of lectures [2], risk communication, biological terrorism, chemical agent terrorism, and the effects of ionizing radiation; legal issues; and special topics sections. The latter were topical lectures from the perspectives of participants' home countries. An ASI core leadership team provided course materials and also stimulated discussions on responses to the current threat of weapons of mass destruction and/or other acts of terrorism from an international perspective. Such interactive discussion among participants was a key component of the ASI. Course co-directors were Drs. Stikova and Rumm.

Structurally, the ASI began with the AMA BDLS course, a detailed introduction, followed by a week of blocks of in-depth instruction and discussion on individual topic areas, plus a series of "special focus sessions" such as the legal perspective of fighting terrorism globally, and a communication section (e.g., risk communication, and the

Supercourse). The ASI also addressed preparedness for special populations during man-made and/or natural disasters.

Synopsis of Topics Covered

Scope of the Threat

There is now a heightened awareness of the threat posed by CBR agents, and of the continued evolution of terrorist operations. Terrorists have used anthrax spores, chemical weapons, conventional explosives, and have intentionally contaminated foods. They now are better organized, well-financed, and have increased access to CBR agents.

Specific Threats

Chemical threats include both lethal and non-lethal agents. Lethal agents such as sarin and VX may be odorless and colorless and exposure can either be via inhalation or via dermal contact. Medical responses, depending on level of exposure, range from sweating and difficulty breathing, on to seizures and death. Response to a chemical event requires protecting medical personnel, removing affected victims from the contaminated environment, decontaminating both victims and the affected area, and medical treatment.

Biological threats can be naturally occurring, as in the case of influenza and SARS, or can be intentional events such as the use of anthrax spores in 2001, or potentially, smallpox. The U.S. Centers for Disease Control and Prevention (CDC) classifies biological agents into three categories according to threat level. Category A agents have the potential for high fatality rate and/or spread, and include *Bacillus anthracis*, Botulinum toxin, *Francisella tularensis*, hemorrhagic fever viruses, *Variola* (smallpox) virus and *Yersinia pestis*. Category B agents are significant public health threats with less potential for mass casualties, and include toxins such as ricin toxin and Staphylococcal enterotoxin B, and infectious agents such as *Brucella* and classic food safety threats such as *Salmonella* and *Shigella*. Category C agents include emerging infectious diseases such as hantavirus, Nipah virus, SARS and avian influenza [3].

Radiological events may result from intentional use or from accidents. Effects of radiation exposure include acute radiation syndrome, internal radionuclide contamination, and the long term effects of organ failure and malignancies. With past radiation events, there has been a very high ratio of psychological effects as compared to physical injuries. Therefore, response to a radiological episode must focus both on medical injuries and psychological injuries, as well as on communication and public education regarding the event.

CBR threats require a response that is “resource intensive” – a heavy investment of both resources and personnel, both to prepare for an attack, and especially to respond to an attack. Even a moderately severe CBR attack could immediately occupy all available hospital beds – intensive care, acute care and psychiatric care beds – in much of a country.

Natural disasters such as the recent earthquake in Pakistan, hurricanes in the USA, and the tsunami in the Indian Ocean require much of the same planning, training and response as that used for CBR agents. Preparedness must address a multitude of potential threats both natural and man-made, and planning efforts can have a synergistic benefit to response to multiple threat areas – the benefits of the "all-hazards approach." In addition, several ASI sessions addressed the risk and effective response to influenza pandemics (such as the current H5N1 threat) and discussed how these correlate to threats of natural disasters and/or WMD.

Actions Needed

Given the range of threats, public health preparedness and response requires expertise and collaboration within and between disciplines and among countries, as well as effective communication between health authorities and other ministries within countries. As was discussed in this ASI, it is paramount to agree upon a framework for resource-sharing, with a command and control structure to use in an emergency or a disaster. It is also critical to conduct exercises of existing plans ("testing the systems") in the form of table-tops, functional drills and full-scale exercises, since these exercises objectively assess response systems to be used during an emergency or disaster, and they identify planning and procedural strengths and areas that require improvement.

Other aspects of public health emergency response are important to include in national and international preparedness plans. Governments should adopt broad approaches to prevention, beyond immediate response to CBR threats. Risk communication is one such area, as it will improve the public response to a disaster. A strong legal framework is necessary to the structure of prevention and response. Inclusion and incorporation of "special needs" populations and mental health perspectives into planning are other such areas.

Preparedness through environmental monitoring and disease surveillance may help minimize the spread of diseases such as avian influenza. However, monitoring and disease surveillance capabilities and standards vary from country to country. Whereas some countries such as Macedonia have implemented sophisticated monitoring systems for infectious diseases, others lack the resources to integrate and implement such systems. Standardization of surveillance systems across allied countries, and cross-cutting coordination of responses that includes law-enforcement and military systems [4] is a top priority.

ASI participants agreed that there should be common civil and military defense programs NATO-wide, organized as a system. This should include: information systems with analytic capabilities, a materiel system that supplies personal protective equipment and stocks of medications and antidotes, detection and analysis capabilities, and decontamination systems.

Education and training is crucial, must be improved, and must be central to NATO's emergency response plans. Scientists and medical personnel at all levels must be trained in CBR threats and in their own emergency response systems. Training should reach all NATO countries, and outcomes should be measured (i.e., in terms of the results of drills and exercises). No single approach to education is adequate, as proper training requires a combination of modalities (among them – lecture, hands-on training, exercises and drills). As well, reaching all NATO countries and Eligible Partner countries will require that all these modalities be used maximally. Finally, since training is "perishable" – its impact fades with time – there must be a system in place for repeated training – a system of continuing education.

The Supercourse can serve a strong role in educating NATO countries, with its network of over 38,000 scientists in 151 countries and its library of over 2700 lectures [2]. This model offers a unique solution to communication barriers among those working in public health readiness and in public health in general. Its use could be expanded in terms of topics and languages included. Its multiple sources, a decentralized broad base of experts [5], are an advantage that NATO can use. Other recent NATO publications have also stated the need for internet-based public health information networks that cut across the often isolated "silos" of academia, government agencies, industry and the military [6]. In this ASI, Linkov et al. [5] propose a network model that would act as a "civil defense" or "neighborhood watch" system. The Supercourse can improve preparedness and public health related communication and training in a cost-effective way.

Summary Points

The ASI stressed the following points and conclusions:

1. CBR agents pose a risk to public health. There is treatment available for victims of CBR agents, and there are systems that can respond to these agents, contrary to popular opinion. Clinicians and public health systems must know that CBR risk is there, and be ready for it.
2. Definitive care of CBR agent injury includes: emergency resuscitation, decontamination, diagnosis, and proper choice of treatment (antidotes, immunizations, antibiotics, blockers and chelators, among the many); some of these must be administered immediately.
3. Psychological effects are likely to outnumber physical injuries, and public health systems must have the available surge capacity for such patients.
4. An international, coordinated system for both public health communication and public health response is indicated. This system should include equipment, supplies, and medicines.
5. A NATO plan, and a system ready for a variety of CBR emergencies, must include risk communication, staffing and supplies – all in place before public health emergencies occur.
6. Education and training in these subject areas require considerable expansion within NATO and its partner nations. Internet-based training, using a variety of sources for material, must be included for this expansion to be effective.

This was an important and timely ASI in light of the global threats of terrorism and disasters. Participants felt that they increased their own personal knowledge base, but just as importantly, learned from each other in an international venue. They worked cooperatively to try to shape potential training and planning solutions against future threats. Such training and collaboration must be continued and we encourage efforts to fund and promote such continued interaction among scientists on an international level.

Recommendations

Given the inevitability of future public health disasters and emergencies and the challenges that NATO Partner countries and NATO Eligible countries face in dealing with them, action is needed to create an effective, unified response system, both within and between partner nations. Participants concurred with the following recommendations that should be implemented and standardized across NATO:

1. Internet-linked communication system – A system, both between NATO countries and within countries, similar to the Health Alert Network in the United States. Such a system would be relatively inexpensive, efficient, and can be readily accessible online. Participants rated this area as the top priority because of its urgency, as well as its probable cost-effectiveness.
2. Infrastructure for natural disaster and CBR agent response – Develop and maintain an infrastructure that identifies and quickly responds to a natural or man-made disaster. This might include a strategic international stockpile, similar to the Strategic National Stockpile (SNS) in the United States.
3. Training – Ongoing training should be planned, organized and targeted to key government and university members in each NATO country and each NATO Eligible country. When doing so, the planning should be incorporated long-term into NATO goals, and there should be a range of approaches to the training. These should include internet-based distance learning and multiple sources such as the Supercourse.

Future NATO-ASI

There was consensus among participants that more ASI's are needed on the combined topics of public health preparedness and CBR agents. For a 2007 ASI, participants suggested that the following be included:

1. The AMA Advanced Disaster Life Support Course (as a follow-on to this year's BDLS);
2. Update reports of national preparedness systems;
3. Two or more days of scientific presentations and didactic material;
4. A full-day preparedness drill/ exercise followed by discussion and debriefing;
5. Committees or working groups that produce and submit to NATO an action plan regarding establishing an internet-based system for communication and distance-learning education in public health preparedness linking NATO and NATO Eligible countries. Logistics of this network and sources of training material such as the Supercourse should be identified.

References

[1] Basic Disaster Life Support Course, American Medical Association, copyright 2004.
[2] Global Health Network Supercourse Project. www.pitt.edu/~super1/.
[3] Department of Health and Human Services, Centers for Disease Control and Prevention, Emergency Preparedness and Response. www.bt.cdc.gov.
[4] Caplinskiene, M. Strategic Management of Mass Disasters Using Forensic Implication Operations. NATO Advanced Study Institute: Strengthening National Public Health Preparedness and Response for Chemical, Biological and Radiological Agent Threats June 19–29, 2006.
[5] Linkov F, LaPorte R, Sauer F, & Shubnikov E. Public Health Preparedness: I-Prevention and Global Health Network Supercourse. NATO Advanced Study Institute: Strengthening National Public Health Preparedness and Response for Chemical, Biological and Radiological Agent Threats. June 19–29, 2006.
[6] Talishinski R, Azmi R, Adlas R, Kedars U, Bakanidze L, Linn S, Rossodivita A, Shishani K, Busmans M, Grabauskas V, Janakasukas D, Shubnikov SE, Trufanov A, Vynograd N, Dorman J, LaPorte R, Linkov F, Noji E, Powell J, Rumm P, Tseytlin E, Volz C. Constructing a NATO Supercourse. NATO Security Through Science Series, D: Information and Communication Security. Volume 5, 2006.

2. American Medical Association – Basic Disaster Life Support (BDLS) Course

Strengthening National Public Health Preparedness and Response to Chemical, Biological and Radiological Agent Threats. Edited by C.E. Cummings and E. Stikova.
IOS Press, 2007.

Healthcare Disasters: *Local Preparedness, Global Response*!

Raymond E. SWIENTON MD, FACEP[a], Italo SUBBARAO DO, MBA[b] and Phillip L. COULE, MD, FACEP[c]
[a] *University of Texas Southwestern Medical Center Dallas, Texas*
[b] *AMA Center for Public Health Preparedness and Disaster Response*
[c] *Center of Operational Medicine Medical College of Georgia, Augusta, Georgia*

Abstract. The world through globalization has become interconnected and economically interdependent. Disasters such as the Indonesian Tsunami, the Pakistan Earthquake, and recent events in the United States of America (USA) such as Hurricanes Katrina and Rita demonstrate the need for an internationally accepted standardized response to disasters. Global public health issues like Severe Acute Respiratory Syndrome (SARS), risk of pandemic influenza as well as acts of terrorism worldwide reinforce the importance of strengthening the international capabilities of disaster response. The American Medical Association (AMA), through the National Disaster Life Support Foundation (NDLSF) is providing leadership in this area by providing a standardized disaster preparedness education and training program that is targeting the wide scope of healthcare providers and disaster response personnel. The international community is being encouraged to collaborate with the NDLSF in establishing global standards for disaster education and training. In the quest to do the greatest good for the greatest number of potential survivors in any disaster, it is imperative that the global healthcare community be able to seamlessly integrate in joint responses. The ability to work together effectively is largely dependant upon our fundamental education and training in disaster preparedness. A review of these topics was presented at the 2006 Advanced Study Institute (ASI) Course: Strengthening National Public Health Preparedness and Response for Chemical, Biological, and Radiological Agents Threats held in Skopje, Republic of Macedonia on 20 June 2006.

Report

The world through globalization has become interconnected and economically interdependent. Disasters such as the Indonesian Tsunami, the Pakistan Earthquake, and recent events in the United States of America (USA) such as Hurricanes Katrina and Rita demonstrate the need for an international accepted standardized response to disasters. Global public health issues like Severe Acute Respiratory Syndrome (SARS), risk of pandemic influenza and acts of terrorism worldwide reinforce the importance of strengthening the international capabilities of disaster response.

The American Medical Association (AMA), through the National Disaster Life Support Foundation (NDLSF) is providing leadership in this area by providing a standardized disaster preparedness education and training program that is targeting the wide scope of healthcare providers and disaster response personnel. The international community is being encouraged to collaborate with the NDLSF in establishing global standards for disaster education and training. In the quest to do the greatest good for the greatest number of potential survivors in any disaster, it is imperative that the global healthcare community be able to seamlessly integrate joint responses. The ability to work together effectively is largely dependant upon our fundamental education and training in disaster preparedness. A review of these topics was presented at the 2006 Advanced Study Institute (ASI) Course: Strengthening National Public Health Preparedness and Response for Chemical, Biological, and Radiological Agents Threats held in Skopje, Republic of Macedonia on 20 June 2006.

The National Disaster Life Support Foundation (NDLSF) was formed by the American Medical Association (AMA) and leading academic institutions in disciplines of public health emergencies and disaster medicine. The University of Texas Southwestern Medical Center at Dallas, the Medical College of Georgia, University of Texas Health Science Center at Houston School of Public Health, and the University of Georgia were the founding members of the NDLSF. The NDLSF has created comprehensive, nationally standardized, mass casualty all-hazards education and training programs for healthcare providers and related non-medical personnel. A measurable improvement in disaster preparedness to a critical mass of the healthcare workforce is the mission. Over 40 NDLS program training centers have been established in the United States alone.

This program includes such AMA Press published texts and courses as Advanced Disaster Life Support (ADLS), Basic Disaster Life Support (BDLS), and Core Disaster Life Support (CDLS) [1–3]. There are several other courses and products under development. A detailed description of the NDLS program courses, target audiences and content can be found by reviewing the published texts and additional information at related websites listed in the reference section [4–7]. The American College of Emergency Physicians (ACEP) has described this program as the "gold standard" in this field [8].

With a foundation in disaster preparedness training and education underway in the USA, an initiative is now formally establishing a consortium of domestic and international stakeholder groups. This is viewed as an important step in meeting the *Local Preparedness, Global Response* readiness that will be fostered through the National Disaster Life Support Education Consortium (NDLSEC). The National Disaster Life Support Educational Consortium (NDLSEC) is a formal entity established by the AMA and NDLSF to build consensus, seamlessly integrate, and improve communication among the key academic institutions, governmental agencies, public and private entities. The NDLSEC is a cohort of national and international stakeholders serving as an advisory board. The focus is on academic excellence, research review and performance improvement of the NDLS disaster education and training programs. The NDLSEC uniquely provides a participatory venue for every medical and surgical specialty society comprising the entire house of medicine under the AMA.

The AMA, NDLSF and NDLSEC highly value the international community's contribution and partnership to improve disaster preparedness and global healthcare response to events. International collaboration is important in achieving an effective and sustainable level of disaster preparedness. The International Disaster Life Support (IDLS) program is establishing a network of education and training centers worldwide. The NDLSF has program activity in several international locations including China, Iraq, Mexico, the Caribbean, Europe, South America and the United Arab Emirates. In the IDLS program, countries or regions develop specific disaster education and training programs unique to local needs while maintaining the essential elements that allow seamless integration of trained healthcare providers and response workers.

IDLS is a positive step towards building international healthcare relationships and reducing the common obstacles such as communication barriers. It is the intention that each country or region will establish its own governing committee which will be represented on the international operations of the NDLSEC. The collaboration of international subject matter experts will facilitate the establishment of peer-reviewed evidence based guidelines for healthcare providers worldwide.

The value of standardized education and training is important, as recent experiences have shown, and is positively impacting several key aspects of disaster preparedness. During Hurricanes Katrina and Rita, NDLSF leadership served many roles, including that of senior medical advisors to the State of Louisiana Department of Health and Hospitals. They established and managed surge capacity healthcare facilities including converting an abandoned K-Mart retail store into a facility for special needs patients. Such innovative, just-in-time applications in healthcare delivery are helping to define new models for disaster healthcare services. Publications, such as *Surge Capacity: Providing Safe Care in Emergencies* by Joint Commission on Accreditation of Healthcare Organizations cite these experiences and discuss their defining role in delivering appropriate healthcare in a sufficiency of care environment [9].

A healthcare disaster exists when the response need exceeds the available resources [2]. The NDLSF leadership's application of the principles they founded in the NDLS education and training programs saved lives during the healthcare disasters of Hurricanes Katrina and Rita. Accounts of their involvement have been published by media sources in the USA as well as healthcare news reports [10–12]. These reports discuss pivotal roles in the evacuation of the New Orleans International Airport during the aftermath of Hurricane Katrina as well as other contributions. Overall, the impact of these historic events is best described by *Local Preparedness, Global Response*!

Healthcare Disasters: *Local Preparedness, Global Response*! Local communities must clearly identify likely disaster threats and be prepared to independently manage the first several days or longer with minimal or no outside aid expectations. Even beyond this time period, local disaster preparedness plans must have realistic expectations concerning the availability of external sources of assistance. For example, disaster preparedness must consider the size and scope of a widespread event and the imposing limitations of finite resources being requested by multiple locations. In the USA, no longer can geographically distant or unaffected regions limit their involvement to sending mutual aid and resources to the local affected areas. Unaffected regions, such as neighboring cities or distant states are aware of the need to locally accept evacuees of distant disaster-affected locations, and to actively engage in providing for evacuees' healthcare needs. Hence, in the USA a "global" response is fostered even during domestic disaster events. For example, non-coastal states that have no geographic threat of hurricanes are now developing an all-hazards disaster plan that includes the consequence management of hurricane victims who are likely to be evacuated to these states in large numbers.

Summary

Ensuring appropriate levels of local healthcare preparedness and global response capabilities requires a solid foundation in standardized disaster preparedness education and training for healthcare providers.

Through globalization the world is greatly interconnected and disaster's affecting one country ultimately affects the world. The AMA and NDLSF are fostering a common language and response to disasters through developing internationally standardized education and training programs such as the IDLS. Through the creation of NDLSEC the AMA is attempting to establish international relationships to build a solid foundation towards global disaster preparedness.

References

[1] Coule P.L., Schwartz R., Swienton R.E. (editors). Advanced Disaster Life Support (ADLS) Provider Manual. Version 1.0, 2.0. AMA Press, Chicago, 2003.
[2] Coule P.L., Dallas C., James J.J., Lillibridge S.R., Pepe P.E., Schwartz R., Swienton R.E. (editors). Basic Disaster Life Support (BDLS) Provider Manual. Version 2.5. AMA Press, Chicago, 2003.
[3] Coule P.L., Fowler R.L., Lakey D., Miller R.G., Swienton R.E. (editors). Core Disaster Life Support (CDLS) Provider Manual. Version 1.5. AMA Press, Chicago, 2004.
[4] http://ndlsf.org/, accessed 17 October 2006.
[5] http://bdls.com/, accessed 17 October 2006.
[6] http://www.ama-assn.org/ama/pub/category/12606.html/, National Disaster Life Support Program. Accessed 17 October 2006.
[7] http://www.ama-assn.org/ama/pub/category/6206.html, AMA Center for Public Health Preparedness and Disaster Response, accessed 17 October 2006.
[8] Schwartz, R., Werlinich, T. Media Advisory: The American College of Emergency Physicians (ACEP) and the National Disaster Life Support Foundation (NDLSF) Combine Efforts to Train Disaster Medical Personnel. ACEP, September 26, 2005.
[9] *Surge Capacity: Providing Safe Care in Emergencies* by Joint Commission on Accreditation of Healthcare Organizations (JCAHO) and Joint Commission Resources, 2006.
[10] The Newshour With Jim Lehrer. PBS. *National. Surge Hospitals: A Look at dealing with the injured and ill from Hurricane Katrina.* Susan Dentzer reporting. 7:00–8:00 PM EST, September 8, 2005.
[11] Silverman, J., Elsevier Global Medical News; Cover Story "We're Going to Need God's Help", ACEP News. American College of Emergency Physicians. October 2005.
[12] Silverman, J., "We're Going to Need God's Help", Surgery News: The Official Newspaper of the American College of Surgeons, October 2005, pp. 1and 4.

3. Epidemiology

Strengthening National Public Health Preparedness and Response to Chemical, Biological and Radiological Agent Threats. Edited by C.E. Cummings and E. Stikova.
IOS Press, 2007.

Pivotal Steps to Building Global Health Security

Elin A. GURSKY
Principal Deputy for Biodefense, ANSER/Analytic Services, Inc.

Abstract. The 21st century has already forecast a continuum of health security challenges that will strain current medical and public health systems. The world must now prepare for an influenza pandemic similar to the 1918 Spanish flu, the scale of which has the potential to cause catastrophic global losses to populations, civil infrastructures and economies. NATO is uniquely positioned to guide and inform regional planning, detection and response efforts to promote global health security. In June 2006, a NATO Advanced Studies Institute (ASI) in Skopje, Macedonia convened 60 participants from 11 countries over the course of two weeks to address strategies to mitigate and contain global public health and health security threats. Four key areas were identified in which NATO's leadership will be vital: convening future meetings of medical and public health experts with government officials to foster collaborative regional disease control efforts; developing health information systems to promote disease surveillance, detection and reporting; facilitating regional training and exercises to create an effective and interoperable workforce; and educating leaders in core issues relating to health security and population protection.

Keywords. NATO, Global health security threats, Population protection, Medical experts, Public health experts, Government officials

Background

Late 20th-century medical, surgical, and pharmacological advances have been significant factors responsible for increasing longevity of life, but they pale by comparison to huge advances wrought from public health. Vaccination against the major childhood diseases; improved sanitation; expanded programs for family planning, maternal reproductive health, and prenatal care; and the application of lifestyle and dietary changes to reduce coronary and cerebral artery incidents are only some of the factors that have produced U.S. decreases in infant and maternal mortality by 90% and 99% respectively.[1] Similar benefits from public health advances have been witnessed outside the United States as well. Infant mortality rates have declined by 50% in 90% of most countries worldwide,[2] and there have been global increases in average life expectancy at birth by almost 20 years.[3] The global use of the inexpensive measles vaccine has prevented almost 3,000 deaths daily from what was once considered the most lethal

[1] "Ten Great Public Health Achievements—United States, 1900–1999," *Morbidity and Mortality Weekly Report*, vol. 48, no. 12, April 2, 1999, pp. 241–243, http://www.cdc.gov/MMWR/preview/mmwrhtml/00056796.htm.

[2] "Remarks by William Foege, M.D., M.P.H., Gates Fellow," World Health Assembly 2000, Geneva, Switzerland, http://www.gatesfoundation.org/MediaCenter/Speeches/GHSpeeches/BFSpeechWHA-000516.htm.

[3] World Health Statistics, World Health Organization, 2006, http://www.who.int/whosis/en/index.html.

agent in the world.[4] From 1999 through 2002, measles deaths decreased by 30% globally.[5] Aggressive vaccination campaigns can also be credited with a greater than 99% reduction in polio cases, from an estimated 350,000 cases in 1988 to just fewer than 700 at the end of 2003.[6] In that same period, polio was eliminated worldwide; it had been endemic in more than 125 countries, but that number fell to just six.[7] The number of neonatal tetanus deaths decreased from 800,000 worldwide in the 1980s to 180,000 in 2002.[8]

Despite these achievements, the new century has illuminated a continuum of health security challenges that will continue to strain medical and public health systems. For example, a cycle of climatic change has instigated a cluster of natural disasters and hazards. These include the European heat wave (2003), the Maharashtra floods (2005), the Kashmir earthquake (2005), Hurricanes Katrina and Rita on the U.S. Gulf Coast (2005), the Indian Ocean earthquake and Southeast Asian tsunami (2005), the Java earthquake (2006), and the southern Leyte mudslides (2006).[9] These events caused death, disability, and population displacement affecting more than 7 million people. Additionally, climate and environmental change and their affects on wind patterns, arthropod habitats, herd movements, and many other related factors have contributed to the emergence and dissemination of new pathogens such as human immunodeficiency virus, West Nile virus, *E. coli* 0157:H7, hantavirus, viral hemorrhagic fevers (such as Marburg, Ebola, dengue, and yellow fevers), highly pathogenic avian influenza, and *Staphylococcus aureus*.[10]

The increasingly global movement of products and people has also challenged systems and strategy designed to protect populations. Airline travel alone accounts for the movement of approximately 500,000 passengers across international borders each day.[11] The global transport of agricultural and animal products provides a venue for the potential introduction of a host of emerging pathogens to all corners of the world. In recent years, outbreaks of diseases such as severe acute respiratory syndrome (SARS), the Ebola virus, and bovine spongiform encephalopathy (mad cow disease) have frightened the public, disrupted global commerce, contributed to large economic losses, and jeopardized diplomatic relations.[12] Over the past few years, foreign agricultural trade of animals and animal products and food and grain products by the United States has

---

[4] "Remarks by William Foege."

[5] World Health Organization, Immunization, Vaccines and Biologicals Department, "Achieving [Millennium Development Goals] With Immunization Successes," August 2004, http://www.who.int/mdg/goals/goal4/050511_immunization_ml.pdf.

[6] "Fact Sheets: Poliomyelitis," Global Polio Eradication Initiative, http://www.polioeradication.org/factsheets.asp.

[7] WHO Immunization, Vaccines and Biologics Department, "Achieving [Millennium Development Goals] With Immunization Successes."

[8] Ibid.

[9] George Pararas-Carayannis, "Climate Changes, Natural and Man-Made Disasters—Assessment of Risks, Preparedness and Mitigation," presented at the Climate Change Disaster Preparedness Conference, Kiev, Ukraine, October 2003.

[10] C.J. Peters, "Hurrying Toward Disaster?" *Perspectives in Health Magazine*, vol. 7, no. 1, 2002, http://www.paho.org/English/DPI/Number14_article3_5.htm.

[11] State of the World Forum, September 2000, http://www.simulconference.com/clients/sowf/dispatches/dispatch2.html.

[12] William B. Karesh and Robert A. Cook, "The Human-Animal Link" *Foreign Affairs,* July/August 2005, http://www.foreignaffairs.org/20050701faessay84403/william-b-karesh-robert-a-cook/the-human-animal-link.html.

steadily increased.[13] In 2005, the United States spent $59.3 billion on foreign agricultural imports, up from $41.9 billion in 2002.[14]

Most of the world's political, health, and agricultural leaders are now preparing for a pandemic that may equal or exceed the effects of the Spanish flu, which killed tens of thousands globally just within its first few months.[15] The 1918 influenza pandemic eventually spread in three waves over the course of a single year to every continent except Antarctica and ultimately claimed 40 million to 100 million lives.[16] The H5N1 avian influenza virus, first identified in 1997 and a possible cause of this potential pandemic,[17] has already spread to 59 countries on three continents[18] and has resulted in the death of 148 people.[19] Aggressive monitoring of poultry has led to culling more than 140 million H5N1 infected birds at a cost in excess of $10 billion.[20]

Deliberate attacks by terrorists likewise threaten the collective confidence in governments to protect their citizens. Since the attacks on the World Trade Center and Pentagon in 2001, the world has witnessed the anthrax letter attacks in the United States (2001) and bombings of nightclubs (Bali, 2002) and mass transportations systems (Madrid, 2004; London, 2005; and Mumbai, 2006).

The opportunity for disease to spread across a variety of vectors—food, animal, and people—both by natural and deliberate means necessitates heightened vigilance and agility of countries and their health systems. Yet the capacity to identify and contain cross-border epidemics or respond to other types of catastrophic health events stretches the capability of most developed nations. For developing and non-developed countries strained by limited resources and infrastructure, the challenge to address these issues is daunting. Proof that these issues are of glaring importance was the inclusion of infectious diseases (along with global energy security and education) as one of three main agenda items in the most recent G8 meeting.[21]

Over the past five decades, the North Atlantic Treaty Organization has been at the heart of efforts to establish new patterns of cooperation and mutual understanding in countries linked by the North Atlantic and has committed itself to essential new activities in the interest of a wider stability by addressing issues such as economic distress, the collapse of political order, and the proliferation of weapons of mass destruction.

---

[13] U.S. Department of Agriculture, Economic Research Service, "Foreign Agricultural Trade of the United States: Monthly Summary," July 2006, http://www.ers.usda.gov/Data/FATUS/MonthlySummary.htm.

[14] Ibid.

[15] Public Broadcasting System, *The American Experience*, "Influenza 1918: People & Events—The First Wave," http://www.pbs.org/wgbh/amex/influenza/peopleevents/pandeAMEX86.html.

[16] Institute of Medicine, Board on Global Health, "The Threat of Pandemic Influenza: Are We Ready?" Workshop Summary (2005), National Academies Press, http://www.nap.edu/books/0309095042/html/7.html.

[17] World Health Organization, "Avian Influenza: Assessing the Pandemic Threat," January 2005, http://www.who.int/csr/disease/influenza/WHO_CDS_2005_29/en/index.html.

[18] World Organisation for Animal Health, "Update on Avian Influenza in Animals (Type H5)," October 4, 2006, http://www.oie.int/downld/AVIAN%20INFLUENZA/A_AI-Asia.htm.

[19] "Cumulative Number of Confirmed Human Cases of Avian Influenza A/(H5N1) Reported to WHO" as of October 3, 2006, http://www.who.int/csr/disease/avian_influenza/country/cases_table_2006_10_03/en/index.html.

[20] "Health, Nutrition and Population in East Asia and Pacific—Economic Impact of Avian Flu: Global Program for Avian Influenza and Human Pandemic," World Bank, http://web.worldbank.org/WBSITE/EXTERNAL/COUNTRIES/EASTASIAPACIFICEXT/EXTEAPREGTOPHEANUT/0,,contentMDK:20713527~pagePK:34004173~piPK:34003707~theSitePK:503048,00.html.

[21] Russian President Vladimir Putin, address to G8 visitors, Saint Petersburg, Russia, 2006, http://en.g8russia.ru/.

Established in 1949 to safeguard the freedom and security of its (now 26) member countries, NATO accepted the challenge of disease and disasters as a threat to regional stability in the closing decade of the 20th century. In 1998, NATO helped establish the Euro-Atlantic Disaster Response Coordination Centre, which is responsible for coordinating the responses of Euro-Atlantic Partnership Council countries to disasters occurring in the Euro-Atlantic Partnership Council area and for maintaining close liaison with the United Nations and the European Union as well as other organizations involved in international disaster response. Since its inauguration, the Euro-Atlantic Disaster Response Coordination Centre has been involved in major operations, including the Ukrainian floods of 1998 and the Kosovo refugee crisis in 1998 and 1999.[22]

In 2003, NATO sponsored an Advanced Research Workshop on "Preparedness against bioterrorism and re-emerging infectious diseases, regional capabilities, needs and expectations in Central and Eastern Europe countries." In 2005, and under NATO auspices, the Center for Biosecurity of the University of Pittsburgh Medical Center, the Center for Transatlantic Relations of the Johns Hopkins University, and the Transatlantic Biosecurity Network created and facilitated a transatlantic bioterrorism exercise dubbed "Atlantic Storm." With distinguished former ministers and ambassadors playing the roles of key leaders of Germany, Netherlands, Sweden, and Turkey, this exercise notionally explored the decisions and responses to a smallpox attack.[23] Among the deliberations were fundamental issues such as whether to share stores of vaccine with countries that had little to none, whether borders should remain open to permit the continued movement of goods and people, and what unifying authority could manage an effective, coordinated response. The exercise illustrated the potential social, economic, and political disruption that would follow an international epidemic, but it illuminated the benefits of advanced preparation among the international medical, public health, and diplomatic communities to minimize death, illness, and financial dislocation.[24]

There is pervasive global consensus that disease presents one of the greatest threats to stability, economy, and security. "There are no islands in the world today," no "domestic diseases and international diseases," Kofi Annan, the Secretary-General of the United Nations, stated in 2001. "We live in a global village. We live in a shrinking world.... no one is isolated, no one can be smug and sit in his or her corner and say, 'I'm safe because it is somewhere else.'"[25]

The global challenge before the public health and medical communities is one that must inspire active efforts within and across countries. In June 2006, NATO sponsored

---

[22] Establishment of the Euro-Atlantic Disaster Response Coordination Centre was endorsed by the Euro-Atlantic Partnership Council ministers on May 29, 1998, and the center was inaugurated on June 3, 1998, jointly by NATO's Secretary General, the Russian ambassador to NATO, and a representative from the UN Office for the Coordination of Humanitarian Affairs. The centre is located in the V building at NATO headquarters. (See "NATO's Role in Disaster Assistance," NATO, 2001, http://www.nato.int/eadrcc/mcda-e.pdf.)

[23] Exercise participants were played by current and former officials from the following countries and organizations, including the presidents of the European Commission, France, and the United States; the Chancellor of Germany; the Prime Ministers of Canada, Germany, Italy, the Netherlands, Poland, Sweden, and the United Kingdom; and the Director General of the World Health Organization. "Atlantic Storm," Center for Biosecurity, University of Pittsburgh Medical Center, http://www.atlantic-storm.org/about.html.

[24] Bradley T. Smith et al., "Navigating the Storm: Report and Recommendations From the *Atlantic Storm* Exercise," *Biosecurity and Bioterrorism: Biodefense Strategy, Practice, and Science*, vol. 3, no. 3, September 2005, pp. 256–267, http://www.liebertonline.com/doi/abs/10.1089/bsp.2005.3.256.

[25] United Nations Secretary-General Kofi Anna, remarks and question-and-answer session with U.S. Chamber of Commerce, June 1, 2001, http://www.un.org/apps/sg/offthecuff.asp?nid=212.

an Advanced Studies Institute titled "Strengthening National Public Health Preparedness and Response for Chemical, Biological, and Radiological Agents Threats."[26] Over the course of two weeks, almost 60 participants from 11 countries met in Skopje, Macedonia, to learn and share their expertise and efforts to address current public health and health security concerns. The following attempts to summarize the lessons learned.

Lesson One: Connecting Critical Parts

In the past 30 years, 30 new or emerging pathogens have been identified around the world.[27] Nearly all emerging infectious disease episodes of the past ten years have been zoonotic.[28] The critical association between animal and human health is now being validated daily by aggressive global preparations to avert a pandemic of avian influenza. Yet despite the well-recognized chain of infection from animal to human host, the integration of health and agriculture programs remains elusive. Discrete funding patterns, missions, and organizational systems all too frequently result in stovepipes that limit the cross-connection of ideas and efforts. Policies may fail to take advantage of specific knowledge sets in neighboring organizations. Professionals may be ignorant of similar initiatives being conducted in sister agencies. At day's end, the loss of opportunities to expand finite budgets and to share colleagues' expertise and perspective results in a loss for all.

Public health and medical practitioners should identify ways to bring their colleagues who work in related agencies into discussion and shared activities for mutual benefit. NATO can support these efforts by hosting cross-agency meetings within countries and within regions. The goals of these meetings should include providing forums in which professionals can discuss strategy that may harmonize their efforts, identify barriers to shared activities, and implement new systems of collaboration.

Lesson Two: Moving and Sharing Critical Information

Disease outbreaks cannot be contained if they are not discovered. Without sophisticated laboratory equipment and disease-reporting systems connecting the continuum of the public health, medical, and laboratory communities, infectious illnesses will implant themselves within populations to spawn new and multiple generations of sickness and death.

In today's globalized and fast-paced environment, information technology is nothing less than a necessity. The *surveillance and detection* of communicable diseases must move at least as fast as the *spread* of the disease in a community.

The use of information technology in disease identification and containment is probably the single most cost-effective tool within the healthcare and health security environments. Yet the implementation of health information systems remains problem-

---

[26] NATO Public Diplomacy Division, ASI Award HSD.982109.

[27] Jeffrey Koplan, "CDC Works in Global Environment," *U.S. Medicine*, January 2001, http://www.usmedicine.com/article.cfm?articleID=136&issueID=20.

[28] Institute of Medicine, Board on Global Health, "The Emergence of Zoonotic Diseases: Understanding the Impact on Animal and Human Health—Workshop Summary," http://www.iom.edu/CMS/3783/3924/4347.aspx.

atic for many reasons. First and foremost is the significant initial investment. However, when best-practice business processes are used to enable rapid collection and a seamless interface with clinical and patient data inputs and subsequently foster rapid situational awareness about hyperendemic disease incidence, costs and risks are significantly reduced. Indeed, numerous studies have demonstrated that, after the first years following implementation of health information systems that promote robust disease management across patients and populations, such systems pay for themselves.[29] Initial estimates suggest that implementing health information technology could lead to annual savings of over $77.8 billion in the United States alone.[30]

In addition to concerns about costs, there are understandably large issues regarding information system requirements (such as "What do we need the system to look like?" and "What do we need it to do?"); the privacy and protection of identified health information; information system oversight, maintenance, and upgrades; and determining which information flows need to go to which organization.

In fact, these issues need not be perceived as overwhelming. Over the past decade, huge advances have been made in developing efficient and reliable information system architectures the scalability of which makes it possible to build incrementally over time, as needs emerge and experience informs. Moreover, cost-sharing across the public and business sectors, as well as across smaller and contiguous countries, will bring investments into line with other key capital initiatives.

NATO should consider bringing together leaders in information technology and health informatics with public health and medical experts to discuss fundamental issues of information system needs, requirements, and architecture and possible implementation approaches. The potential impact of H5N1 avian influenza on the European community speaks to the urgency of this endeavor.

Lesson Three: Building an Effective and Resilient Workforce

Emerging pathogens and rapid advances in biotechnology pose a steep and continuous learning curve for medical, public health, and related professionals. New techniques in detection, protection, and response abound in the literature, erupt from industry, and appear on the Internet. For professionals working 10 to 14 or more hours a day, it is profoundly difficult to keep up with this bombardment of information.

Not discounting the importance of passive didactics, including reading and self-study forms of continuing education, and even one- or two-day meetings and workshops, one of the single most important means of honing skills and informing decision making is through the use of tabletop and full-scale exercises. Few other forms of learning are as successful at bringing together professionals from across different responder sectors. This strategy improves team-building skills and joint approaches to assessing the rapid flows of situational information. It forces familiarity and harmony

---

[29] "Health and Human Services' Estimate of Health Care Cost Savings Resulting From the Use of Information Technology," letter from David A. Powner, Government Accountability Office Director of Information Technology Management Issues, to Rep. Jim Nussle, chairman of the U.S. House of Representatives Budget Committee, February 16, 2005, GAO-05-309R, http://www.gao.gov/new.items/d05309r.pdf.

[30] Center for Information Technology Leadership, *HIEI: Healthcare Information Exchange & Interoperability*, 2004, http://www.citl.org/research/HIEI.htm.

of effort and minimizes the use of sector-specific jargon that may impede swift and effective decision making.

Exercises have many costs associated with them. To create, build, and facilitate scenarios that accurately depict disease and catastrophic events requires medical and public health acumen and a cadre of other experts. Participation in one- and two-day exercises once or twice a year means that the supply of frontline professionals in the clinics and communities is depleted and must be backfilled by a workforce that is not resident. It also requires pulling from the lines not only the most skilled professionals to participate in these exercises, but also the support staff—less-skilled workers, hospital security guards, telephone operators, truck drivers, and others—who will play a vital role during a pandemic or similar event causing illness and social disruption.

Building region-wide, ongoing training and exercise frameworks for its member countries is a worthy objective. NATO sponsorship will assure the highest quality of exercise and tabletop development and will legitimize the importance of country participation in these events.

Lesson Four: Educating World Leaders

None of the lessons presented above will move forward unless they are accepted at the highest levels of each country's government. Yet, with the wide and overwhelming menu of competing priorities and problems facing our leaders today, it is often difficult to bring health issues to the forefront. One of the most daunting challenges for medical and public health professionals is how to guide and support government leaders in implementing the strategies that will maximize available resources to protect the health of their people, as well as to ensure the economies and stability of their countries.

Two key issues are at the core of this dilemma: knowledge and access. Most people in high leadership slots are not positioned because of their expertise in the physical sciences. Yet, their ability to make good medical and public health decisions requires understanding of such concepts as "transmissible/communicable disease," "incubation period," "exposure," "vaccine efficacy," and "isolation." As such it will necessitate trust in and dependence upon health ministers, physicians, epidemiologists and others who possess this scientific expertise.

One solution for improving knowledge and access that can foster effective decision making is to develop executive leadership courses. For example, NATO could establish and support four two-day executive courses in offsite locations during the course of a year. Each year, these executive courses would bring together eight of the highest-level political and health leaders in three to four geographically contiguous countries (with a maximum of 30 participants) for two days of intensive learning and discussion. Mini-scenarios would test and challenge decision making, using and balancing both science and greater political goods to help generate the most practical decisions to inform policy and action. In addition to NATO's Advanced Studies Institutes, other models are the Aspen Institute's Business and Society Program, which (in November 2005) convened a group of leaders addressing the theme of "Developing Leaders for a Sustainable Global Society,"[31] and the Harvard Macy Institute's executive course for European

[31] The Aspen Institute's Business and Society Program has been bringing together leadership development experts from corporations, academic institutions, and professional service firms from around the world to share perspectives about the role education can play in contributing to social and environmental well being.

healthcare leaders to focus on driving change—a six-day program designed to assist healthcare leaders in what they should be doing to advance healthcare.[32] Through these and other programs yet to be developed, the fundamental and inextricable connectivity between the health of a country's people and the vitality of a country's economy can be enhanced and expanded.

Closing Thoughts

Naturally occurring pandemics and terrorist attacks pose large health security challenges in this new century. Global concerns of pandemic flu are but one example of the interdependency of approaches and efforts to protect the world's peoples.

Organizations such as NATO are uniquely positioned to provide a unifying and guiding force across numbers of countries that cannot, in and of themselves, support a standalone, effective disease-combating infrastructure. Moreover, even countries that possess a greater span of resources recognize that their independent efforts will not provide a sufficient aggregation of capabilities to halt a pandemic. Connecting countries in collaborative efforts and educating world leaders on how to make effective decisions to protect the health of their populations are essential steps in ensuring global health security.

Medical and public health leaders must help bridge government efforts to identify and contain disease outbreaks. With NATO's leadership and vision, these efforts can become effective strategies for global stability, economy, and health.

The author would like to thank Ms. Sweta Batni for her research assistance with this article.

Dialogues with executive educators bring together leaders to share expertise and insights about their role in education to develop leaders to reflect on the complex financial, social, and environmental challenges they may face in today's global society and look to build and support a learning network of practitioners. Aspen Institute, "Dialogues with Executive Educators," 1999–2004, http://www.aspeninstitute.org/site/c.huLWJeMRKpH/b.730759/k.43EE/Dialogues_with_Executive_Educators.htm.

[32] See "Executive Course for European Health Care Leaders to Focus on Driving Change," Harvard Macy Institute, *HMI World*, May/June 2004, http://www.hmiworld.org/hmi/issues/May_June_2004/macy.html.

Improvement of an Integrated System of Disease Surveillance in Georgia Under International Cooperation

Lela BAKANIDZE, Paata IMNADZE, Shota TSANAVA and Nikoloz TSERTSVADZE
National Center for Disease Control and Medical Statistics of Georgia, Tbilisi, Georgia

Abstract. Effective communicable disease control relies on efficient, high-quality disease surveillance. A well-functioning disease surveillance system provides information for planning, implementation, monitoring, and evaluation of public health programs. The National Center for Disease Control and Medical Statistics of Georgia carries out surveillance on communicable and non-communicable diseases (including especially dangerous pathogens). The chain of notification and reporting of diseases in the country is established. All institutions and providers rendering health care services to the population must notify the local public health service whenever they diagnose, suspect, or even receive positive laboratory results. US DoD Defense Threat Reduction Agency (DTRA) has started implementation of a project to improve surveillance systems (standardized and repeatable disease monitoring systems, mobile epidemiological response teams and secure transportation of infectious agents), improve communications and information technology (including electronic communicable disease reporting system), improve biosafety and physical security of central reference laboratories and the safe transportation of pathogens, and collaborate on establishing new national rules and regulations relating to BWPPP.

Introduction

The Georgian National Health Policy, adopted in 1999, declared the reduction of communicable and socially dangerous diseases to be a major priority for maintaining and improving the health of the Georgian population over the next decade.

Effective communicable disease control relies on functioning high-quality disease surveillance. This requires the systematic and regular collection of information on the occurrence, distribution, and trends of an event on an ongoing basis with sufficient accuracy and completeness to provide the basis for action. A well-functioning disease surveillance system provides information for planning, implementation, monitoring, and evaluation of public health programs. It includes case detection and registration, case confirmation, data reporting, data analysis, outbreak investigation, response and preparedness activities, feedback, and communication. Health authorities must provide appropriate supervision, training, and resources to assure that surveillance systems operate properly.

1. Surveillance System in Georgia

An ideal surveillance system is sensitive enough to correctly identify all cases of a particular disease occurring in the community; particularly for especially dangerous pathogens; any single case must be detected.

All clinically diagnosed or laboratory confirmed cases of communicable diseases that come to health facilities for treatment or consultation (irrespective of whether they are reported urgently or once a month) must be registered.

The usefulness of public health surveillance data depends on its uniformity, simplicity, and timeliness. State and local public health officials use the information about occurrence of diseases to accurately monitor trends, plan, institute policies and decisions, and evaluate the effectiveness of interventions.

Established surveillance systems should be regularly reviewed on the basis of explicit criteria of usefulness, cost and quality; systems should be modified as a result of such review. Attributes of quality include: (i) sensitivity; (ii) specificity; (iii) representativeness; (iv) timeliness; (v) simplicity; (vi) flexibility; and (vii) acceptability. Often evaluation of surveillance systems is limited in scope and content. The sensitivity of a surveillance system is its ability to detect health events (completeness of reporting). Its specificity is inversely proportional to the number of false positive reports. Representativeness can be measured by comparing surveillance data covering part of the population to either nationwide data, where available, or to random sample-survey data. Simplicity in a system means it is easy to understand and implement, and is therefore usually relatively cheap and flexible. A flexible system is easily adapted by adding new notifiable diseases or conditions or extending it to additional population groups. Acceptability depends on perceived public health importance of the event under surveillance, recognition of individual contributions and time required for reports.

All notifiable diseases and conditions are divided into two groups according to their implications to public health surveillance and response:

- Diseases and conditions of which health authorities must be notified urgently;
- Diseases and conditions of which health authorities must be notified monthly.

All reportable diseases and conditions are divided into two groups as well:

- Cases of infectious diseases and conditions subject to monthly reporting;
- Cases of infectious diseases and conditions subject to annual reporting.

Groups of notifiable and reportable diseases do not match.

Immediately and without any delay, information should be submitted for the following internationally regulated especially dangerous infections: plague, cholera, yellow fever, poliomyelitis, viral hemorrhagic fevers, tularemia, anthrax, rabies, SARS smallpox, tick-borne encephalitis, influenza caused by a new subtype. Urgent notification must be done for group of cases of any infectious disease (excluding acute respiratory infections and influenza).

The National Center for Disease Control and Medical Statistics of Georgia (NCDC) was founded on the basis of Georgian Station for Plague Control in 1996, NCDC is an integral part of the Georgian Public Health system, it reports to the Ministry of Labor, Health and Social Affairs of Georgia. The main responsibilities include: Surveillance on communicable and non-communicable diseases; control (planning and implementation) of important (Public Health) diseases; instituting preventive measures, promoting healthy lifestyle; and gathering and processing medical statistical data. NCDC owns Georgian National collection of especially dangerous pathogens.

One of the main responsibilities of NCDC is surveillance on communicable and non-communicable diseases (including especially dangerous pathogens). All institutions and providers rendering health care services to the population, regardless of their subordination and forms of ownership, including laboratories and private care providers, must notify the local public health service whenever they diagnose, suspect, or even receive positive laboratory results. The NCDC determines and annually updates the list of notifiable and reportable diseases on the basis of the current epidemiological situation.

Surveillance Network

NCDC

↑

Regional PH Centers – 11

↑

District (rayon) PH Center – 66

↑

Medical Facilities and Doctors

Figure 1. Organization of the Surveillance Network in Georgia.

After the collapse of the Soviet Union, Georgia experienced severe economic difficulty and possessed a shattered health care system. Health care and surveillance of communicable and non-communicable diseases operated only through international cooperation and with the support of international organizations, like USAID, AIHA, WHO, UNICEF, UNFPA, Global Fund, GAVI, VRF, etc. With their help, NCDC succeeded in controlling and preventing important (Public Health) diseases; by expanding their health promotion activities, administering immunization programs, and gathering and processing medical statistical data. During the 1990's there were several particular achievements such as enhancing situational analysis, instituting public health strategy design, piloting several guidelines (CD surveillance guidelines for CPH and health care providers, these are polyclinics and so-called FAPs – feldsher – ambulatory points at the lowest level of the figure job aids for rayon (district)-level CPH and facility workers, etc. were worked out); strengthening human resource capacity, increasing training and supervision, developing and implementing nationwide policy (endorsed by MoLHSA Decrees), and issuing communicable disease surveillance guidelines. Additionally, standard case definitions were determined for AFP/Polio, measles, diphtheria, mumps, rubella/CRS, pertussis, tetanus, acute viral hepatitis, rabies, shigellosis, salmonellosis, cholera, bacterial meningitis, influenza H5N1, upcoming case definitions for anthrax, plague, tularemia, brucellosis, TBE, HFRS, CCHF, etc.

It should be noted that NCDC has broad research and scientific potential. About 60% of the staff are specialists with a university education, 32 of whom have scientific degrees (candidates and doctors of sciences) and research projects, funded by US DOD/DTRA, DHHS/BTEP, DOE/IPP, CDC, ISTC, STCU, SIDA, JAICA, etc. The list of implemented and ongoing projects, given in the Table 1, demonstrates such efforts:

Table 1. List of Selected Projects, Implemented and Ongoing at NCDC Georgia

Name of the Project	Dates	Donors
International Training and Research in Emerging Infectious Diseases	1997–2002	Fogarty Center, NIH
Establishing Epidemiological Network on the Territory of Georgia	1997	"Open Society Georgia" Foundation
Improvement of Epidemiological Network in Georgia	1998	"Open Society Georgia" Foundation
Reproductive Health Survey	1999–2000	UNFPA, UNICEF, USAID, UNHCR, AIHA, CDC
Nutritional Status of Children Under Five Years of Age in Six Regions of Georgia	2000–2001	USAID/Save the Children-US, Georgia Field Office
Provision of Epidemiological Survey Services on Baku – Tbilisi – Ceyhan Pipeline Route	2003	British Petroleum
International Training and Research in Emerging Infectious Diseases (ITREID)		Fogarty Center, NIH,
Enhanced Epidemiologic and Laboratory Diagnostic Capacity for the Control of Botulinum Intoxication in Georgia	2001–2004	US DHHS/BTEP; Collaborator: CDC, Atlanta, GA, USA
Prevention of Amebiasis and Creation of Diagnostic Test-Systems for *E. histolytica Strains* Isolated in Georgia	2001–2005	US DHHS/BTEP; Collaborators: University of Virginia Health System, Charlottesville VA, USA
Molecular Epidemiology and Antibiotic-Resistance of Bacterial Infections in Georgia	2002–2005	US DHHS/BTEP; Collaborator: University of Maryland / School of Medicine / Baltimore, MD, USA
Clinical and Molecular Epidemiology of Drug-Resistant Tuberculosis in the Republic of Georgia and the Caucasus	2002–2005	US DHHS/BTEP; Collaborator: Emory University, Atlanta, GA, USA
Epidemiology, Molecular Characteristics and Clinical Course of HCV Infection in Georgia	2002–2005	US DHHS/BTEP; Collaborator: Johns Hopkins University, Baltimore, Maryland, USA
Application of Molecular Fingerprinting to Geographical Characterization and Epidemiological Surveillance of Natural foci of *Yersinia pestis* and *Francisella tularensis* in the Republic of Georgia	2004	US DHS, Collaborator: Lawrence Livermore National Laboratories, Livermore, CA, USA
Ecology, Genetic Clustering, and Virulence of *Yersinia pestis* Strains Isolated from Natural Foci of Plague in Georgia	2005–2008	US DoD DTRA
Application of Molecular Fingerprinting to Geographical Characterization and Epidemiological Surveillance of Natural foci of *Yersinia pestis* and *Francisella tularensis* in the Republic of Georgia	2006	US DHS, Collaborator: Lawrence Livermore National Laboratories, Livermore, CA, USA

US DoD Defense Threat Reduction Agency (DTRA) has started implementation of a project to improve surveillance systems (standardized and repeatable disease monitoring systems, mobile epidemiological response teams and secure transportation of infectious agents), improve communications and information technology (including electronic communicable disease reporting system), improve biosafety and physical security of central reference laboratories and the safe transportation of pathogens, and collaborate on establishing new national rules and regulations relating to BWPPP. All these elements are supplied with verifiable training in their field.

The Central Epidemiological Monitoring Station (EMS) was instituted at NCDC. It is equipped with modern sophisticated equipment, like light cyclers, RT – PCR, etc. EMS carries out surveillance on especially dangerous infections. Here first avian flu case in Georgia was identified.

2. Conclusions

This type of project implementation is beneficial for Georgia. Increasing the biosecurity and biosafety at biological facilities and improving disease surveillance infrastructure and capabilities through state of the art technology will result in timely detection and response to outbreaks or epidemics. Early warning systems linked to data collection and disease surveillance systems will benefit Georgia by reducing disease transmission and risk. Additionally, the improved capacity of NCDC reference laboratories provides the opportunity to engage in a spectrum of collaborative research.

References

[1] Surveillance and Control of Communicable Diseases: Guidelines for Public Health Services in Georgia. Abt. Associates Inc. 2005.

[2] Thacker S.B., Parrish R.G., Trowbridge F.L. World Health Stat Q, 1988; 41 (1):11–8.

4. Informatics

Strengthening National Public Health Preparedness and Response to Chemical, Biological and Radiological Agent Threats. Edited by C.E. Cummings and E. Stikova.
IOS Press, 2007.

Public Health Preparedness: I-Prevention and Global Health Network Supercourse

Faina LINKOV PhD[a], Ronald LAPORTE PhD[a], Francois SAUER, MD[a] and Eugene SHUBNIKOV, MD[b]
[a]*Graduate School of Public Health, University of Pittsburgh, Pittsburgh, Pennsylvania*
[b]*Siberian (Novosibirsk, Russia) and Pittsburgh (PA, USA) Supercourse Centers*

Abstract. Public health systems in the US and many other countries are poorly prepared for severe (*Actually we handle the majority well!*) natural and man made disasters. Teachers and public health educators across the country and worldwide have only limited materials to educate their students on the risk and risk factors of West Nile virus, avian flu, bioterrorism, and other disasters. The Global Network Supercourse group provides a model that can help to exchange educational materials rapidly and at low cost. The Supercourse group *(Do you capitalize "group?")* developed a large number of scientific lectures on public health preparedness and disasters, and made them available through the Supercourse network of over 38,000 scientists 151 countries. This article describes a novel approach of; **i-Prevention**, the application of the Internet to prevent bioterrorism and other threats. Additionally, the paper outlines the concept of community watch and its applicability to the modern era of the Internet technologies. A neighborhood watch is where people in communities ban together to watch over each other, the neighbors, and communities are deterrents against crime. The principles of the neighborhood watch can be established on the Internet, with a Global Health network neighborhood watch to act as a deterrent against bioterrorism and other disaster, and mitigate the damage should an attack occur.

Keywords. Public Health Education, disaster preparedness

Introduction

A few kilograms of anthrax spores have the potential to kill most of the people in London, Moscow, or Los Angeles. Effective public health systems are critical for the prevention and management of such a potential catastrophe and other disasters [1]. However, public health systems in the U.S. and many other countries are simply unprepared for the majority of natural and man made disasters. The West Nile Virus (WNV) outbreak and the threat of avian influenza in the past several years demonstrated the vulnerability of our homeland to attacks by biologic agents. Confusion, error, miscommunication and misdiagnosis helped create an "epidemic" of fear [2]. From January 2003 to the end of October 2003, 44 states and the District of Columbia reported more than 7,700 human cases of WNV infection, resulting in 166 deaths [3]. Luckily, the disease was not highly virulent. If it were, thousands of people would have died. A bioterrorist would use more deadly agents, and s/he would spread it over much greater areas creating extraordinary havoc. These threats demonstrate that, public health systems are ill-prepared to handle bioterrorist attacks.

The governments of the U.S., the United Kingdom (U.K.) and others annually spend hundreds of millions of dollars to support their public health infrastructures and to thwart bioterrorism. However, even with the increase in funding, the WNV episode, the events of Sept 11th 2001, the threat of avian flu, and other episodes demonstrate how poorly prepared our public health institutions are for an attack. We argue that improvement will not occur by pouring more money into established paradigms of public health. Rather new, more effective technologies need to be tested and introduced. A progressive and new approach is **i-Prevention**, the application of the Internet to prevent bioterrorism.

Supercourse Background and i-Prevention

The term i-Prevention was coined by the Global Health Network Supercourse group at the University of Pittsburgh. Supercourse is a library of over 2600 PowerPoint lectures on prevention, shared by over 38,000 members of the network from 151 countries. We have developed a simple strategy to develop and distribute lectures based on e-business models. The concept is "lecture shareware" where scientists place their most exciting lectures into a "library of lectures" for widespread dissemination on the Internet. This is a free library where any teacher can use any lecture. It is necessary to maintain high quality of the lectures in the Supercourse library, which is why Supercourse developers are using quality assurance systems from industry such as those of Deming [4]. Instead of 50 students seeing our "golden" lectures each year, as many as 40,000 could do so. Long after we retire or die, our lectures can aid future generations of faculty; "great" lectures will be immortalized in virtual libraries. Supercourse could improve the teaching of faculty members world-wide, and will lead to a better education and training for future generations.

Epidemiologists, preparedness professionals, and other faculty members today rarely disseminate and share lectures, especially in the area of public health preparedness and disaster mitigation. We do not build upon other lectures and we rarely obtain feedback from other faculty members on our lectures or courses. This weakness is shared by all science. In addition, existing university distance education courses are rigidly copyrighted, and too expensive for developing countries. This process leads to little self-improvement as there is virtually no scientific approach to quality control in lectures. Why cannot we approach transnational epidemiology training as an open manufacturing system? Why cannot we act as global faculty, make our lectures available to an Internet lecture library and let the global scientific faculty examine, debug, and improve our top lectures, also saving us a lot of work?

Supercourse Model

The Supercourse model offers a unique solution to communication barriers among those working in the field public health preparedness and public health in general. We have constructed "freeware" lectures on epidemiology, the Internet and global health for medical schools, nursing schools, veterinary schools, dental schools, etc., worldwide [5]. Use of the lectures by faculty is free, and comment is welcome. Supercourse group with core members in Pittsburgh as well as around the globe, developed a technology for inexpensive, sustainable global training. The Global Health Network Supercourse project consists of the following:

1. **Open Source:** A Global faculty develops and shares their best, most passionate lectures in the area of prevention and the Internet using an open source model. This benefits all. Experienced faculty can beef up their lectures; new instructors reduce preparation time and have better lectures; faculty members and educators in developing countries can freely access current prevention information.
2. **Statistical Quality Assurance:** We have established a Deming Model of statistical quality control to monitor lectures over time.
3. **"Support" Educators:** The Library of Lectures consists of exciting lectures by academic prevention experts in the field. The classroom teacher "signs out" lectures for free, like borrowing a library book. We "coach" the teacher rather than directly teaching students from a distance.
4. **Text books:** The British Medical Association put text books on line for us.
5. **Multilingual:** For global use, the first lecture is available in 8 languages. A large number of Supercourse lectures are available in multiple languages, including Spanish, Russian, Chinese, French, and others. We are experimenting with machine translation as well.
6. **Faculty:** Fifteen Noble Prize winners, the U.S. Surgeon General, 60 Institute of Medicine (IOM) members, the head of the National Institutes of Health (NIH), the former head of CDC, and other top level scientists contributed lectures.
7. **JIT lectures:** Within days after a disaster, scientific lectures are provided, e.g., regarding the Bam Earthquake, the Indian Ocean tsunami, and Hurricanes in the U.S.
8. **Mirrored Servers and CDs:** We have 45 mirrored servers in Egypt, Sudan, China, Mongolia and others. We have distributed 20,000 Supercourse CDs.

The Supercourse model is very effective for sharing information in the field of bioterrorism and other natural and man made disasters. Bioterrorism is unlike most other forms of morbidity, as the conditions, exposure, agents, and populations attacked are virtually infinite. No one could have predicted the importance of "dead crows" that indicate the spread of West Nile Fever [6]. It is extremely difficult to predict and guard against a bioterrorist attack, as there are too many targets, too many means to penetrate the targets, the attacks are too rare and the bioterrorists are too wily. Pouring more money into the current system is like shoring up the Maginot line, as the cunning enemy easily slips around our staid defenses. We need to develop a broader, "flatter" and nimbler organization to combat bioterrorism, with thousands more eyes, as the existing hierarchical public health system is not by itself sufficiently agile to compete with bioterrorists.

Internet Civil Defense

The military defense of the U.S., U.K., and other countries at war has always been partly the responsibility of our citizens. We would argue that this is also the case for the war against bioterrorism, where citizens should be the first, and most important line of defense. We have the opportunity to build an i-Prevention system by interconnecting citizens on the Internet to form an Internet Civil Defense [7].

Civil Defense/Home Guard during WWII included individuals trained to prevent battle damage within a country. Civil defense represented the identification of leaders in the community who would oversee activities during air raids. These were "trusted agents" who would guide people to shelters, or help to dig people out from ruined buildings. Citizens knew that when they heard the alarm that the leaders would be a trusted source of information. During the Cold War of the 1950's and 1960's Civil Defense became more coordinated; mitigation of panic was an essential component. The Civil Defense leaders were trained as to where to take people for protection, and to watch over the neighborhoods to prevent looting.

Towards the end of the cold war civil defense/home guard started to wane. A primary reason was that the arsenals on both sides became so destructive that a war would almost yield mutual annihilation. Any attempt at civil defense would have little effect in mitigating Armageddon. However, with the fall of Communism the threat of complete annihilation became dimmer. Yet, focal disasters, whether man-made (terrorism, nuclear reactors), or infectious (bioterrorism, emerging disease), continue to plague citizens. These will be rare, unusual and virtually impossible to predict.

As Civil Defense waned, a new form of home guard grew in response to an internal enemy, rising crime rates and resulting fear – Neighborhood Watch. A Neighborhood Watch is where concerned citizens ban together to watch over each other, the neighbors, and communities are deterrents against crime. Why not establish the same principle for bioterrorism, whereby neighbors watch out for neighbors and for the unusual events, but do so on the Internet? The goal of the Global Health Network Neighborhood Watch will be to act as a deterrent against bioterrorism, and mitigate the damage in case of an attack. The beauty of such a system is that with the Internet, there is a death to distance. Thus infectious disease experts from Egypt could be consulted over the Internet within minutes after the diagnosis of WNV. Moreover, in the community of 300,000, over 200,000 will be connected to the Internet, especially in the developed world. Community leaders could thereby provide updated information to those neighbors who are not connected, to alleviate fear and to guide people to safety.

Instead of building an inflexible Maginot Line defense as we are now, perhaps we should consider an ever alert, flexible electronic matrix civil defense as our first line of defense against bioterrorism. This public health system could collect information concerning possible bioterrorism activity locally, which can be filtered with the goal of being Information Dominant over bioterrorism. Suspicious behaviors can be investigated, "nipping in the bud" planned bioterrorism activities. This was beautifully encapsulated in a Nature editorial: "But the panic of an ill-informed public, as the society suggests, could be at least as dangerous as an attack itself" [8].

The concept of this system has many advantages. First, it can increase the number of people world-wide on the outlook for bioterrorism from the current 2000 public health experts to 20,000,000 citizens being on Internet Bioterrorism watch. With this, 40,000,000 "eyeballs" will be on the look out for bioterrorism activities. Second, should a major bioterrorism event take place, more people would likely be killed because of panic (the War of the Worlds effect), than because of the terrorism as noted above in the Nature Bioterrorism editorial. Nothing better alleviates fear than having your brother, or neighbor, or friend from Nottingham to come over and say that everything is all right, as this is what s/he had heard from the Civil Defense Network portal. People believe information from trusted agents more than from people we do not know.

Thus the Civil Defense team acts as a secure information pipeline to filter accurate information through a trusted matrix of people. Finally, in the days and weeks following the outbreak, this network can continue to deliver accurate information to people in need.

The concept of a neighborhood watch can be more specific, like a school watch. Children are well suited to be monitored for bioterrorism. They are some of the most susceptible, they transmit agents readily at school and to family, and they are some of the easiest to monitor. In many ways, they are like the "canary in the coal mine", the earliest indication of an attack. However, there is virtually no monitoring of children in either the U.S. or U.K. for bioterrorism. Bringing a school watch onto the Internet could guard our children with a local to global Internet School watch.

Neighborhood Watch and Networking

How would we grow an Internet Civil Defense Network? This would be quite simple, and has been done to a smaller extent with the Global Health Network (www.pitt.edu/~super1/). The global health network has over 38000 academic experts in prevention and the Internet. The goal is to identify a core group who have web access, and who want to protect their communities against bioterrorism. Certification systems could evolve, similar to what was used during WWII. We could use the "chain-letter" approach to recruit, sending email letters to people who we know would participate, and they would then send it to 5 of their friends, and onward. The concept is simple; it is also one of what is called 6 degrees of separation, where we are separated by any in the world by no more than 6 relationships. Thus, if you were asked to contact a Nigerian Physician, you could do so by going through at most 6 people. If constructed correctly, potentially every person on the Internet could be identified and linked – for the protection of us, our families, and our Internet neighbors around the world.

Ideally we would reach at least one person in each local community. This person would then set up the model for the Internet Neighborhood Watch by doing the exact same procedure a with traditional Neighborhood watch, but over the Internet. Local area Civil Defense watch will combine with other census tracts, and other countries, as the watch quickly branches upward to become a wider area global watch.

Conclusions

Current public health approaches towards prevention of bioterrorism are inadequate, have too few eyes, and are too costly. Also, little is done concerning the panic that would follow an attack. Information sharing through the Global Health Network Supercourse project offers a model to deliver the information on public health preparedness to scientists worldwide. Bringing the eyes, minds and ears of citizenry in the U.S., the U.K. and the World to fight bioterrorism offers a unique opportunity to thwart and mitigate bioterrorism.

Our academic network would be invaluable for establishing common services, preparedness plans, a public health network, immediate assessment and capacity building. There is the need for social responsibility for intercommunication, cooperation and coordination between academy, industry, NGOs and governments at times of crisis.

Oftentimes in the time of disasters, experts of academia and industry are not used. By engaging the best minds locally and globally in many disciplines at the time of a disaster, we can become better and better at creating relevant knowledge and translating this knowledge into "economically feasible present actions" for crisis management. Our network may become the largest global health network in the world. The Civil Defense/Home Guard model has proven to be effective in our countries and fits very well with the key ideas of the Supercourse project. An Internet Civil Defense Model as first line of defense would have a major effect to thwart and mitigate bioterrorism.

References

[1] Rosen P, Coping with bioterrorism *BMJ* 2000; 320:71–72. Available from http://www.bmj.com/cgi/content/full/320/7227/71.

[2] LaPorte RE, Ronan A, Sauer F, Saad R, Shubnikov E. 2002 June. Bioterrorism and the "epidemiology of fear." Lancet Infect Dis 2(6):326.

[3] US Food and Drug Administration http://www.fda.gov/fdac/features/2003/103_virus.html accessed May 20, 2006.

[4] Gabor A. The Man who Discovered Quality. How W. Edwards Deming brought the quality revolution to America – the stories of ford, Xerox, and GM. Random House 1990.

[5] Global Health Network Supercourse Project www.pitt.edu/~super1 accessed May 20, 2006.

[6] Steinhauer, J, Miller J. In New York Outbreak, Glimpse of Gaps in Biological Defenses: New York Times, Oct. 11, 1999.

[7] LaPorte RE, Sauer F, Dearwater S, Aaron DJ, Sekikawa A, Sa ES, Shubnikov E. Towards an Internet Civil Defense against Bioterrorism {Personal View}. Lancet Infectious Disease 2001.

[8] Editorial. A health warning on bioterrorism. Nature; 2000; 13:109.

Public Health Preparedness and Effective Access to Information: Getting the Most Out of Your PC

Eugene TSEYTLIN, MS
Department of Biomedical Informatics, University of Pittsburgh School of Medicine, Pittsburgh, Pennsylvania

Abstract. The field of public health preparedness encompasses many important areas, including preparedness programs in responding to bioterrorism, mass casualties, chemical emergencies, natural disasters, radiation emergencies, and infectious disease outbreaks. The strength of these programs lies in its shared mission to strengthen public health training and workforce development, especially in the areas with limited access to scientific literature and the Internet. The majority of public health emergencies can be prevented and/or mitigated with improved access to scientific literature, open access software, and open access educational modules. Getting information to the right people at the right time is essential. Today, access to information usually implies access to a reliable computer. Contrary to some myths, one does not need the most expensive computer and fastest Internet connection in order to access and utilize high quality educational materials. Computers that often get disposed in U.S. and other developed countries can be used just as effectively as the latest and greatest machines for access to the information. The key to improved access to information is a functional computer. The key to functional computer is oftentimes not hardware, but its software, configuration, user awareness, and attitude towards computer usage. This chapter will discuss how to protect and tune a computer for optimal performance without needing to buy a new one. It will also outline how older machines could be effectively used to access public health preparedness information online.

Keywords. Public Health Preparedness, public health informatics, education

Introduction

Today, the field of public health is facing many challenges, including bioterrorism, natural disasters, and emerging infections. Public health informatics, defined as the systematic application of information and computer science and technology to public health practice, research, and learning, is the emerging discipline that integrates public health and information technology [1]. Major challenges include developing coherent, integrated national public health information systems, increasing integration efforts between public health and clinical care systems, and addressing pervasive concerns about the effects of information technology on confidentiality and privacy [2].

Dramatic changes and technological advances that took place over the course of the 20th century coincided with a major transformation in the way in which we think about the information and information sharing. The last decade introduced to the public the concept of the Internet, which provided professors, doctors, and the general public the opportunity to exchange the information much more efficiently than ever before. Internet revolutionized the way we thought about information. Suddenly, it became possible to share large volumes of information between continents immediately and with minimal expenses. Professors from various universities, including medical and public health schools, have improved access to variety of latest research developments through the Internet. Such valuable sources of information include search engines, free electronic journals, open source lectures, etc. Over the last several years, PubMed Central, BioMed Central, and the Public Library of Science have joined the slightly older PubMed.

Advances in the mode of electronic data exchange did not necessarily translate into better access to public health preparedness information, both in the developing and developed countries. The information exchange in the area of public health preparedness remains inadequate due to problems such as digital divide, poor transfer of information between research lab and classroom, lack of adequate educational and curricula tailored to the needs of students from various educational backgrounds. This paper will discuss the major barriers to the public health preparedness information exchange in the U.S. and globally. It will also outline several initiatives that have a great potential to resolve some of the information exchange deficiencies that exist today.

Intelligent Tutoring Systems: Potential Tool for Improved Public Health Preparedness Education

The notion of intelligent machines for teaching purposes can be traced back to 1926 when Sidney L. Pressey built a machine with multiple choice questions and answers. This machine delivered questions and provided immediate feedback to the user. Educational psychologists have since reported that carefully designed individualized tutoring produces the best learning for most people [3].

Broadly defined, an intelligent tutoring system is educational software containing an artificial intelligence component. The software tracks students' work, tailoring feedback and hints along the way. By collecting information on a particular student's performance, the software can make inferences about strengths and weaknesses, and can suggest additional work [4].

Intelligent Tutoring Systems (ITS) can be quite effective in teaching fields that are difficult to master and have few qualified educators. For example, in the field of pathology, two ITS System, SlideTutor and ReportTutor, were tested and yielded 60% and 160% learning gains respectively [5].

Similar systems could be created in the field of public health preparedness education. Intelligent Tutoring Systems in general are still in research phase, however, once they mature, there is limitless potential in their deployment. The use of ITS will be most valuable in the geographic regions where qualified educators are scarce.

Identification of Barriers to the Information Sharing in Public Health Preparedness: Digital Divide and Information Sharing Examples from Cuba

Despite the emergence of so many efficient ways to share the electronic data, problems remain with the exchange of health information in the area of public health informatics, due, in part, to the digital divide. Digital divide separates those who have access to latest technology (Internet, for example) and those who do not have it. There is lack information sharing especially between developing and developed world with the often forgotten contribution of researchers in developing nations to the world's corpus of information on medical science [8]. Less than 15% of the world has access to the Internet and the majority of those who are connected come from developed countries [9].

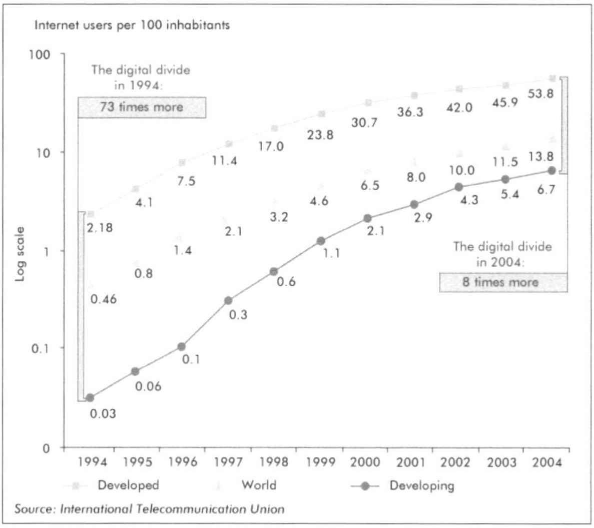

Figure 1. Internet usage in developed vs. developing world [9].

One might argue that the more resources the government pumps into public health preparedness system, the better the outcomes would be should natural or man made disaster occur. However, this situation is not always true. Preparedness is not always the function of economy or GNP per capita. Most developing countries experienced a dramatic growth (from 40 to 63 years) in their population life expectancy in the last decades of the 20th century. Most of these changes can be attributed not to the advances in clinical medicine, but due to better knowledge of hygiene practice or better information sharing.

A very good model of public health preparedness education can be found in Cuba, an island country with a relatively small per capita income. Life expectancy in Cuba is very similar to the life expectancy in the developed world, whereas annual incomes of Cubans and expenditures on healthcare are fractions of those in U.S. or U.K. Despite hurricanes and other natural disasters, the number of natural disaster victims in Cuba remains very low. Cuba has an effective community based public health preparedness system. There, public health preparedness is taught by community members, not government officials. Ultimately, good information exchange and primary prevention approaches save more lives than discoveries in the area of clinical medicine. This information can be exchanged and shared even with older computers.

Clearly, information on public health preparedness is not being adequately shared between the developing and developed world. If one decides to use PubMed, one of the most popular resources for finding health research references, he or she will get plenty of references on preparedness in U.S. or U.K., but very few references on health issues in the developing world. Most materials from local journals (published in languages other than English) will not be available for the researchers outside that specific institution or that country. Lack of PubMed entries about public health preparedness in the developing world does not mean that such materials do not exist. The materials are available, but not being shared.

Getting Public Health Preparedness Information Through the Digital Divide: One Laptop per Child Project

> *"Children will be able to learn by doing, not just through instruction—they will be able to open up new fronts for their education, particularly peer-to-peer learning."*
> —Kofi Annan.

Educating children is a very important component for public health preparedness. In times of crises, children are one of the most vulnerable groups. They are also the group who can identify the warning signs of natural and man made disasters and notify the adults about them. Sadly, public health preparedness is not taught to any great extent at the grade school level and even at the college level. Children on the other side of digital divide have even more difficulties getting the information on public health preparedness. A $100 laptop project is a new initiative that has a great potential to provide children with important educational tools. One Laptop per Child (OLPC) is a non-profit association dedicated to research to develop a $100 laptop—a technology that could revolutionize how we educate the world's children [6]. The $100 laptop is an education project that originated at MIT. It aims to create an inexpensive laptop computer intended to provide every child in the world access to knowledge and modern forms of education. Public health preparedness could become an important component of this educational media.

Figure 2. $100 laptop [6].

This laptop is specifically designed for developing nations. It has a ragged design, as few moving components as possible, bright colors and even a hand crank for times when electricity is not available. Ad-hoc wireless mesh networking may be used to allow many machines Internet access from one connection. The pricing goal is currently expected to start at around US$135–140 and will slowly decline to $100. People in developed nations can purchase those laptops at triple price, so that two children would get free laptops.

Due to the high cost of copyrighted software, the $100 computers will be based on open source software. Open source software refers to computer software available with its source code and under an open source license. Such a license permits anyone to study, change, and improve the software, and to distribute the unmodified or modified software. It is the most prominent example of open source development [11].

The laptops will be sold to governments and issued to children by schools on a basis of one laptop per child, however, markets could be extended to reach children throughout the world.

This great project is another way to bridge the digital divide and provide children in developing nations access to educational materials and other information.

A variety of educational and training programs to address public health informatics knowledge and skills are urgently needed by the public health workforce. These programs must be tailored to meet varying needs, ranging from basic information for the entire public health workforce to more specialized, in-depth management skills for public health managers [7].

Getting Public Health Preparedness Information Through the Digital Divide: Getting the Most Out of Your PC

Another way to bridge the digital divide is to recycle old computers that otherwise would end up in landfills in developed nations. Organizations renew their computer hardware once every 3 years. Sometimes replaced computers get donated to charity or resold at surplus stores. Often times they get disposed off when computer is percieved to be too old or if the organization just doesn't want to deal with hassles of recycling it. Consumers will typically discard their computer hardware because it is the simplest thing to do for them.

Computers are rarely disposed off because they are broken, they are usually thrown out when they seem to be too old. Old computers seem slow in comparison to the newer models. But the basic functionality is still there. While few today will settle for an eight year old PC, it might be a godsend to people in developing nations.

The reason older computers are slow has nothing to do with its hardware, but more with its software. With time, users tend to upgrade their software which becomes more resource demanding and computer hardware simply can't handle it anymore. If user runs Microsoft Windows 2000 on a PC that was designed for Windows 98, he or she will not be satisfied with computer performance. Coupled with that is malware that computer might be infected with and the fact that Microsoft Windows OS performance tend to degrade with time. Given those reasons computer slows down to a crawl and typically gets replaced.

Old computers could be retrofitted with open source software that was customized for older machines and shipped to developing countries. Those machines could be just as effectively used for information access as their newer cousins.

Organizations that already have computers can extend their usable life span and protect them from hackers and malware by following these steps:

1) use anti-virus software
2) never open executable binary attachments!
3) frequently update your system and virus definitions
4) uninstall software you do not need
5) disable unneeded services and programs from starting at startup
6) do not upgrade software unless there is a reason to do so
7) do not login with administrator rights

8) occasionally check your system for AdWare (free tools are available)
9) occasionally "defrag" your hard-drive (use built in windows utility for that)

When these procedures are followed, organizations that do not have resources to replace their hardware will get more out of their machines.

While this may not solve all the problems of the digital divide, it may close the gap somewhat.

Future steps

What could be done to improve information sharing in the area of public health preparedness? Universal access to the latest research literature and Internet technologies would be the ideal solution, but it is not realistic in the near future. Perhaps instead of trying to achieve universal Internet/computer access, countries should view the Internet as a backbone linking to networks of inexpensive hubs. We can close the divide by bringing the Internet to knowledge distribution hubs in Nairobi where intelligent agents search the information, and then content is delivered using traditional means of information distribution already in place, such as teaching, libraries, fax and photo copies, reproduction in local publications, lectures, and word of mouth [10].

References

[1] Yasnoff WA, Overhage JM, Humphreys B, LaVenture M. A National Agenda for Public Health Informatics Summarized Recommendations from the 2001 AMIA Spring Congress J Am Med Inform Assoc. 2001 Nov–Dec; 8(6): 535–545.
[2] Koo D, O'Carroll PW, LaVenture M. Public Health 101 for Informaticians. *J Am Med Inform Assoc.* 2001; 8(6):585–597.
[3] The Encyclopedia of Educational Technology (accessed June 21 2006) http://coe.sdsu.edu/eet/Articles/tutoringsystem/start.htm.
[4] American Association for Artificial Intelligence (accessed June 21 2006) http://www.aaai.org/AITopics/html/tutor.html.
[5] Crowley RS, Medvedeva O, Legowski E, Tseytlin E, Chavan G, Jukic D. Formative evaluation of SlideTutor – an intelligent tutoring system in pathology. APIII 2004, Pittsburgh, PA October 6–8, 2004.
[6] MIT media laboratory (accessed June 21, 2006) http://laptop.media.mit.edu/.
[7] Yasnoff WA, Overhage JM, Humphreys B, LaVenture M. A National Agenda for Public Health Informatics Summarized Recommendations from the 2001 AMIA Spring Congress J Am Med Inform Assoc. 2001 Nov–Dec; 8(6): 535–545.
[8] Gibbs WW. Lost science in the third world. *Sci Am* 1995; August: 76–83.0.
[9] World Summit on the information Society (accessed on June 22, 2006) http://www.itu.int/wsis/tunis/newsroom/stats/.
[10] Ze Y, Shoubnikov E, Acosta B, Sekikawa A, Furuse N, Lee V. Leapfrogging the Digital Divide {eLetter}. Brit Med J Apr. 2001.
[11] Open Source article from Wikipedia, the free encyclopedia www.wikipedia.org accessed June 21, 2006.

Strengthening National Public Health Preparedness and Response to Chemical, Biological and Radiological Agent Threats. Edited by C.E. Cummings and E. Stikova.
IOS Press, 2007.

Information Security Approaches to Provide Social System Continuity in Conditions of Chemical, Biological, Radiological and Nuclear Threats

Ron LAPORTE[a] and Andrey TRUFANOV[b]
[a]*Disease Monitoring and Telecommunications, WHO Collaborating Center, University of Pittsburgh, Pittsburgh, USA*
[b]*Irkutsk State Technical University, Irkutsk, RF*

Abstract. Essential Information Security (InfoSec) problems are discussed and common myths formulated, including that InfoSec development must be identical everywhere; that every entity has to protect its own assets; that confidentiality is the most important element of InfoSec; that terrorists are the only and most crucial adversaries technical tools provide infuse; that mathematics is nothing but cryptography; that the goal is to spend as little on security as possible and that it would be nice if security paid for itself; that standards are void and complicated; that IT team should create the InfoSec system; and that outsourcing is the solution as they know everything. InfoSec myths also apply to the Emergency Preparedness and Response (EPR) field. EPR could better implement Information Security practice and vice versa, stimulate InfoSec development and thus enrich each other. For both, InfoSec and EPR, strong international, cross-agency and interdisciplinary approaches with focusing on information sharing instruments are needed to counteract chemical, biological, radiological and nuclear threats.

Keywords. Information Security, Myths, Emergency Preparedness and Response, chemical, biological, radiological and nuclear threats

Introduction

Information Security (InfoSec) is critical to the strengthening of national public health preparedness and response for chemical, biological, and radiological (CBRN) agent threats. Preparedness for natural and man-made disaster should not be considered separately from information and communication issues. Interconnection is double-sided, that is:

- Information flows usually are damaged when CBRN threats has been activated (traditionally demonstrated in panic), and an information attack might precede and provoke CBRN actions.
- Information Security experience might be developed to the field of Emergency Preparedness and Response (EPR).

Thus, theoretical and practical InfoSec experience might help counteract chemical, biological, and radiological threats for each of these two sides.

First, CBRN threats threaten information exchange. As such, here are some important tips [1] to consider: "During an emergency, tune to a local radio station to obtain information and instructions from emergency officials. Do as they advise and stay away from the disaster scene. Be prepared to relocate if you are advised to do so and follow all instructions carefully."

During a disaster, huge volumes of data, vast distances and lack of reliable information are observed. A DHS report [2] assesses the strengths and weaknesses of the information technology that the Emergency Preparedness and Response Directorate uses to support incident response and recovery operations.

Second, studies of adaptability of theoretical and practical InfoSec experience for counteracting CBRN threats might be of some value. It is reasonable to assess public health – interpretation principles of InfoSec construction if these are applicable to EPR problem.

Information Security – Some Myths and Points

Information security is data protection, as well as a technology problem. There are syntactic, semantic and pragmatic definitions of the term "information". Broadly defined, it involves the protection of various forms of data, services, systems and electronic communications from risk, using appropriate measures. Experts emphasize that security is a people—not technology—problem [3]. Also, experts prefer to consider InfoSec as a process rather than a problem or task. The U.S. National Information Systems Security (InfoSec) Glossary [4] defines Information systems security (INFOSEC) as "the protection of information systems against unauthorized access to or modification of information, whether in storage, processing or transit, and against the denial of service to authorized users or the provision of service to unauthorized users, including those measures necessary to detect, document, and counter such threats."

Our definition of the term implies actors of information processes; that is, "A state and a process of information balance which is beneficial for a proprietor of information resources." InfoSec is about people and data protection is about records. While contemporary technical levels increase drastically, human consciousness lags behind. Thus, InfoSec is moving from the technology industry to the social and economic disciplines.

Myth #1 – All InfoSec Development Must Be Identical

The next step to assess the InfoSec problem is to clarify proprietary aspects of information resources. A common myth is that InfoSec development has to be identical everywhere. In fact, it is paramount to set whose prosperity the information or information system supports. One knows that interests are different for different parties: international, national, corporate or individual.

What should a proprietor do if an employee does not want to accept reliable information that is incongruent with his/her model of information activity/policy? Policies, fundamental approaches, techniques, schemes, and applied InfoSec solutions are essentially different for two entity classes:

- Structures of clear administration hierarchy (Private Investment Corp, for example);

Figure 1. Supercourse, funded by the National Library of Medicine (NIH), is an Internet-based library of lectures on prevention and public health, shared for free by 10,000 members from 151 countries in the Global Health Network.

- Organizations and projects with complicated proprietary and management structures (International Research Project).

Thus, InfoSec is a process and state of information provision of sustainable entity continuity, beneficial to its proprietor. (Information policy defines which process is beneficial and which one is not.) It is not easy to adjust interests of diverse proprietors while gathering, analyzing, interpreting, distributing and applying geographical information [5,6].

We may emphasize that definition of Information Resource (IR) proprietary is one of key factors for the entire InfoSec process.

The following are organizations and projects with complicated propriety and management structures that we have had experience in seeking InfoSec solutions for them:

- Region as a territory which combines groups with common interests and common threats [7];
- Irkutsk State Technical University [8];
- Simulation in radiation physics [9];
- Town building Cadastre of Irkutsk region [6];
- «Supercourse» [10] – international project on public health, epidemiology and the Internet (Fig. 1).

The difficulties we faced are explained by the following:

- Weakness or lack of adequate Infosec policy, professional education, comprehensive theory, legal base, interdisciplinary and international cooperation attitude and skill.
- Existence of multi-proprietor cases contrary to traditional one-proprietor consideration, InfoSec myths, globalization, interdisciplinary misunderstanding/ contradictions, contradictions between information generators and security professionals, diversity of contradictions – international, national, federal, re-

gional, municipal, corporation and individual levels, distrust that communities view each other with.

Myth #2 – Every Entity Has to Protect Its Own Assets

One should not confuse a proprietor of information assets and a manager. Protection of government, corporation and individual information assets may be a goal not only for the proprietors but for other national or international entities. This leads us to the second myth that every entity can and has to protect its own assets by itself and that national InfoSec preparedness system is similar to that of corporation. It is evident that government, corporations and individuals like to protect their own information assets. Also, international organizations, government, NGOs, corporations, and individuals might be engaged to provide national InfoSec. Individual preparedness is an important element of the U.S. strategy for homeland security [11]. So it is necessary to emphasize that it is different to own information resources and to provide InfoSec.

Myth #3 – Confidentiality Is the Most Important Element of InfoSec

Concerning the definition of InfoSec, three widely accepted attributes of information need to be considered: confidentiality, integrity and availability (termed the "CIA"). This triad attempts to ask the question "Which one is of greater value? The answer leads to the myth that confidentiality is the most important element of information security. Under this myth, only hazards to information confidentiality are threats. In previous totalitarian regimes, this was the case, but the first goal of modern information security has become to ensure that information resources are predictably dependable in the face of all sorts of threats, including denial of service attacks. We know that information security policy defines which state is beneficial for a proprietor of information resources (IR) and which is not. So which property is moved to the top of this priority list depends on the policy. Security threats include everything from natural disasters to human error and equipment failure to theft, fraud, vandalism, sabotage, and terrorism. Consequences of these threats bring difficulties, problems and risk to proprietor information assets or employee information activities:

$$\text{RISK} = \frac{\text{Threat X Vulnerability}}{\text{Countermeasures}} \text{ X Value}$$

Myth #4 – Terrorists Are the Only and Most Crucial Adversaries

Within InfoSec, another myth is that terrorists are the only and most crucial adversaries. "If you are a member of my tribe," for instance, "you can be trusted. But if you are not, you should be viewed with suspicion." InfoSec experts believe that insiders' misuse of information, intentional or accidental, poses the greatest information security threat. Thus, we may divide risk that insiders (i) and outsiders (o) pose by threats in different ways:

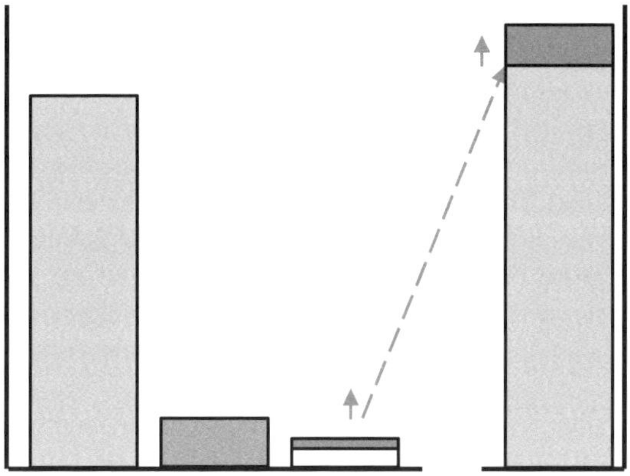

Figure 2. Return on Security Investment (ROSI) illustration.

$$\text{RISK} = \left\{ \frac{\text{Threat}_i \ \text{X} \ \text{Vulnerability}_i}{\text{Countermeasures}_i} + \frac{\text{Threat}_o \ \text{X} \ \text{Vulnerability}_o}{\text{Countermeasures}_o} \right\} \text{X Value}$$

Myth #5 – Technical Tools Provide Infuse; Mathematics Is Nothing But Cryptography

InfoSec theory classifies measures as ethical + legal + organizational + technical + physical + math. However, among experts, there is the myth that technical tools are the only ones to provide InfoSec, and that mathematics is nothing but cryptography. In fact, many threats to information are outside the technical aspects of information systems. The scope needs to be defined broadly enough to expressly and meaningfully encompass the whole of an information system. This includes organizational and individual behaviour, manual aspects of the overall system, as well as aspects of the system supported by computing and communications facilities. In the last the decade, the mathematical theory, net theory, greatly advanced the field of InfoSec practice [12]. Net Theory is about connecting the dots. For example, Stanley Milgram asserted that any two Americans are connected by a mere six intermediaries or "degrees of separation" (Small world nets). Similarly, Albert-Laszlo Barabasi found that any two unrelated web pages are separated by only 19 links.

There are no limits to the growth of the number of links for a node (scale free nets). By analyzing network structure, it provides information on potential vulnerabilities, threats and adversaries. Thus, it makes sense to expand the circle of conventional measure to economical and social ones.

Myth #6 – Spend Minimally in Security

Economical issues raise a very important question; that is, how does one justify spending money on security? At the same time, embedded in this question, is the myth that the goal is to spend as little on security as possible and that it would be nice if security paid for itself. Too much security is a waste, but not enough can pose problems. The challenge is to find that balance between the two. Return on Security Investment (ROSI) [13] may be an answer of great value (Fig. 2). Some examples fortify that role of economical issue have been increased. Thus, a new science of economics of information security emerged. The 5th Conference on the Economics of Information Security (WEIS 2006) was held in Cambridge, U.K in June 2006 [14].

Also, InfoSec classifies the principles of Infosec System Creation in the following manner: system approach; measure diversity; continuity; adequacy; flexibility; simplicity; and openness. It is difficult to define as a myth, but the fact is, there is a common delusion that some principles may be omitted while doing Information Security. As a result, all failures in practice are explained with disregard of the principles. Projects of innovation structure require extra attention and offer the additional principles such as cross-disciplinary, balancing of diversity, standardizing and competition, accountability, and specificity.

Myth #7 – Standards Are Void and Complicated

Instructions and materials are of great value to InfoSec, but the problem is: which of those are trustworthy? That standards are void and complicated is only a myth.

Actually, ISO 17799 [15] is a vivid document comprised of the best and most trustworthy approaches to the practice InfoSec. The 2005 version of the standards contains the following 12 main sections:

- Risk assessment and treatment
- Security policy
- Organization of information security
- Asset management
- Human resources security
- Physical and environmental security
- Communications and operations management
- Access control
- Information systems acquisition, development and maintenance
- Information security incident management
- Business continuity management
- Compliance

Even if the entity is not to be certified under 17799, the latter may be a mark to strive for.

Myth #8 – Outsourcing of InfoSec

The last myth is concerned with those who are responsible for the construction of the InfoSec system. Most experts believe that the IT team should create the InfoSec system and that outsourcing is the solution as they know everything. Ideally, if the InfoSec system is the product of a collaborative effort, proprietors risk mostly. In fact, employers set InfoSec policy and outsourcing brings trustworthy templates. Employees have specific knowledge. Also, cross-silo architecture strengthens construction. In any case, choice and attendance must be accountable and competition between organizations and people should be implemented. It is nonsense if one subject (as IT personnel) orders, develops, implements, performs, controls and measures. Experience shows that there is a definite level of trust to official institutions and government.
(Thus, CIO 2005 Survey clarified that:

- Information security executives have a negative perception of the Department of Homeland Security;
- The color-coded alert system has proved useless.)

Summary

Emergency preparedness and response could better implement InfoSec practice and vice versa, stimulate InfoSec development and thus enrich each other. All the myths and points are applicable to information security in EPR and/or to EPR. The issue of preparedness and response definition for CBRN threats [16] is of value as "information security" is to the InfoSec field. The most sensitive and ill developed issue for both spheres is metrics, which provides a tool to compare events and processes in diverse components of the fields. For both InfoSec and EPR, strong international, cross-agency and interdisciplinary approaches are needed. The focus of this effort could be information sharing as an instrument for the cross-disciplinary collaboration. Below are the tiers that comprise, manage and sustain the knowledge sharing process alive:

1. Contents (Value of Knowledge, Reliability, Trustworthy; Discipline or Science Development Level)
2. Technology (Instruments of Info Exchange)
3. Organization (Schemes of Information Flows, Agency, Corporative- or Project-Rules, Procedures, Instructions, Codes of Practice)
4. Law Basement (National and International Law, Law Enforcement and Practice)
5. Ethics (National and International Aspects)
6. Policy (Corporation, Agency and National)
7. Human Resources (Education, Training and Drilling Level)
8. Personality (Leader Features, Member Attitudes)

It is a bit different to Information Sharing Process which is usually concentrated on 3 principle components 2, 3, 7-Technology, Organization and Human Resources.

References

[1] Emergency Preparedness. http://www.lambtononline.com/emergency_preparedness.

[2] Emergency Preparedness and Response Could Better Integrate Information Technology with Incident Response and Recovery. Department of Homeland Security. Office of Inspector General. Office of Information Technology. OIG-05-36.Sept.2005.

[3] Experts repeat: Security is a people—not technology—problem. http://www.gcn.com/online/vol1_no1/21439-1.html.

[4] National Information Systems Security (Infosec) Glossary http://security.isu.edu/pdf/4009.pdf.

[5] Emergency Preparedness Digest. http://ww3.psepc-sppcc.gc.ca/ep/ep_digest/jm_2000_fea_e.asp.

[6] O. Gluhov, E. Protasova, A. Troufanov. Conflicting Structures in the System of Federal Town-Planning Cadastre: Irkutsk Region. Siberian Regional GIS Conference. Jul 31 – Aug 3, 2002. In Russian. http://www.geomarket.ru/5747.html.

[7] A. Troufanov. Security of Regional actors within of Global Information Processes. 4th Russian Conference "Information Security of Russia in Conditions of Global Information Society. June 24, 2002, http://www.infoforum.ru/detail.php?pagedetail=445.

[8] G. Maltseva, N. Nagaeva, A. Troufanov. Confidentiality in Contemporary Practice of Russian Education. Security of Information Technologies. 2000, N1, pp. 90–93.

[9] A. Petrov, A. Troufanov. Computer Simulation of Radiation Processes and Researcher's Strategy on the Information Market. Irkutsk, 2000. p. 120.
[10] Global Health Network Supercourse: http://www.pitt.edu/~super1/.
[11] Lynn Davis, Tom LaTourrette, David E. Mosher, Lois M. Davis, David R. Howell. Individual Preparedness and Response to Chemical, Radiological, Nuclear, and Biological Terrorist Attacks. MR-1731 RAND, 2003 http://www.rand.org/pubs/monograph_reports/MR1731/index.html.
[12] http://www.nytimes.com/2006/03/12/magazine/312wwln_essay.html?pagewanted=all.
[13] Finally, a Real Return on Security Spending. Feb. 15, 2002 Issue of CIO Magazine http://www.cio.com/archive/021502/security.html.
[14] Workshop on the Economics of Information Security 2006 http://weis2006.econinfosec.org.
[15] ISO 17799 2005 Translated into Plain English. Introduction http://praxiom.com/iso-17799-intro.htm.
[16] What is Emergency Preparedness? http://www.theimo.com/imoweb/EmergencyPrep/WhatIsEP.pdf.

Strengthening National Public Health Preparedness and Response to Chemical, Biological and Radiological Agent Threats. Edited by C.E. Cummings and E. Stikova.
IOS Press, 2007.

The Role of Information Technologies and Science in the Prevention of Bioterrorism

Eugene SHUBNIKOV, MD, Faina LINKOV, PhD, Ronald LAPORTE, PhD and Supercourse Faculty (www.pitt.edu/~super1/faculty/lecturers.htm)
Siberian (Novosibirsk, Russia) and Pittsburgh (PA, USA) Supercourse Centers

Abstract. Education is paramount to preventing all forms of diseases and conditions. Supercourse is a project designed to create a free lecture library of PowerPoint prevention lectures. To date, over 38000 academic faculty from approximately 150 countries with over 2600 available free PowerPoint lectures have participated to integrate Internet-based education into prevention of all forms of diseases and terrorism/bioterrorism. Biological agents are the oldest of the nuclear, biological, and chemical (NBC) triad and have been used by governments in warfare for over 2,500 years [1]. The objective of this chapter is to outline some scientific facts about bioterrorism, to understand terrorism and to help prevent bioterrorism. More specifically, our aim is to: (1) provide information about bioterrorism on the Supercourse collection of lectures; (2) show that terrorism/bioterrorism has had a long history; (3) demonstrate that bioterrorism has occurred worldwide, including in the countries of the Former Soviet Union (FSU); and (4) to conclude that terrorism/bioterrorism, while terrifying, is rare. However, we must be concerned, not paralyzed by fear, and be ready to fight all forms of terrorism/bioterrorism.

Keywords. Bioterrorism, Public Health Education, disaster preparedness

Introduction

What can we do to improve prevention efforts for the populations of our countries, public health persons, non-military, clinicians or veterinarians? The Supercourse team asserts that the Internet provides a powerful and inexpensive tool for the improvement of health and for the protection from bioterrorism. Much has been done in this area already. Supercourse has created a Disaster web page where lectures on terrorism/bioterrorism are available from experts in this area (www.pitt.edu/~super1/disasters/disasters.htm). Fred Muwanga, Supercourse author, defines bioterrorism as "the threat or use of biological agents by individuals or groups motivated by political, religious, ecological, social or for other ideological objectives to inculcate fear or cause illness or death in order to achieve their objective" [2]. Similarly, Rashid Chotani, another Supercourse author, defines bioterrorism as "the unlawful use, or threatened use, of microorganisms or toxins derived from living organisms to produce death or disease in humans, animals, or plants. The act is intended to create fear and/or intimidate governments or societies in pursuit of political, religious, or ideological goals" [3].

Bioterrorism Basics

What makes the use of biological agents so appealing to the terrorist? [3]

- Ease of Acquisition
  - Information is readily accessible on World Wide Web
  - Other sources
- Ease and Economy of Production
  - Only basic microbiology equipment is necessary
  - Small labs require no special licensing
  - Investment to cause 50% casualty rate per sq. km: Conventional weapon $2000, nuclear $800, anthrax $1
- Lethality
  - 50 kg aerosolized anthrax = 100,000 mortality
  - Sverdlovsk experience, former USSR
- Stability
- Infectivity
  - Weaponized agents may be easily spread
  - Clinical symptoms may appear days or weeks after release
- Low Visibility
- Ease and Stealth of Delivery
  - Remote, delayed, undetectable release
  - Difficult/impossible to trace origin of agent

What Can We Do As Medical Professionals?

- Maintain a high index of suspicion by including biological agents in differential diagnoses.
- Learn to recognize historical and physical examination findings suggestive of an exposure to a biological agent.
- Stay informed of local, regional and national epidemiologic trends.
- Be knowledgeable about treatment and prophylaxis of patients exposed to biological agents.
- Know whom to report suspected biological agent exposures and illnesses (Police, State Intelligence agency, Infectious Disease Specialists, Local and State Health Officials) [3].

History of Bioterrorism

Is bioterrorism something new? [4]. The history of biological warfare can be traced back to B.C.

- 6th Century B.C. – Assyrians poisoned the wells of their enemies with rye ergot [3].
- 6th Century B.C. – Solon of Athens poisoned the water supply with hellebore (skunk cabbage), an herb purgative, during the siege of Krissa.

- 184 B.C. – Hannibal's forces hurled earthen pots filled with serpents upon enemy.

More recent bioterrorist events include the following [2]:

- The Polish Resistance: An official from the Polish resistance organization claimed to have killed 200 Germans by using biological agents during World War II. No details of the planning and execution are available, however, nor has there been an official confirmation of this report.
- 1952 – The Mau Mau, an independence movement in Kenya comprised of soldiers from Eastern Africa countries, used a plant toxin derived from an African milk bush to poison livestock. The use of arsenic was also attempted. They intended to cause a direct economic loss and create terror among the rural population. This would result in loss of public faith thereby motivating people to wage war against the British colonialists.
- 1966 – Dr. Suzuki, a Japanese physician and bacteriologist with extensive experience in laboratory science, injected patients and healthcare providers with *Salmonella typhi*. Two hundred developed typhoid and dysentery, and four died. The motive was revenge – he was angry about the treatment he was receiving as a resident in his medical training. However, there are suggestions that he wanted to create clinical cases for his academic research into *Salmonella typhi*.
- 1981 – The Dark Harvest Group protested against the testing of an anthrax bomb on Gruinard Island by the British during World War II and the continued anthrax contamination. The group collected anthrax contaminated soil from the Island and discarded it on the grounds of Britain's biological and chemical weapons research center in Porton Down.
- 1984 – The Rajneeshees used a home made brew of poisonous *salmonella typhimurium* and secretly spread it on fruits and vegetables in salad bars, blue cheese dressing and table-top coffee creamers.
- 1995 – The AUM Shrinkyo attempted to develop weapons using B.anthracis, botulinum toxin, *C.burnetii* and Ebola. It is reported that they tried nine times to disseminate biological agents without success. They used sarin, an organophosphate nerve gas, during Tokyo subway attack [2].

Responses to Bioterrorism

Bioterrorism can be approached from several different levels. Public health responses begin at the finding of the first case. A complimentary view is that of law enforcement where prevention begins before any event takes place, in the 5 year period of preparation. This could be as simple as identifying people who want to take flying lessons but who do not want to learn how to land, or those purchasing Anthrax from web sites. Both of these approaches have to go hand and hand to prevent attacks.

The problem is that few recognize the time course of terrorism. There is a long, protracted time period of planning, followed by a short execution phase and short diagnosis phase. The September 11, 2001 terrorists prepared for almost 5 years before finally executing their attack. The time between execution and first death is typically small. For example, if a nuclear devise were to be detonated, death would ensue within seconds. We want to be able to intervene to prevent children and adults from experiencing a terrorist attack, and death.

International Biological Weapons Documents

- 1925 Geneva Protocol
- 1972 Biological Weapons Convention (signed by 103 nations)
- 1975 Geneva Conventions Ratified

Geneva Protocol: "That the High Contracting Parties, so far as they are not already Parties to Treaties prohibiting such use, accept this prohibition, agree to extend this prohibition to the use of bacteriological methods of warfare and agree to be bound as between themselves according to the terms of this declaration."

Convention on the Prohibition of the Development, Production and Stockpiling of Bacteriological (Biological) and Toxin Weapons and on Their Destruction Biological and chemical weapons have generally been associated with each other in the public mind, and the extensive use of poison gas in World War I (resulting in over a million casualties and over 100,000 deaths) led to the Geneva Protocol of 1925 prohibiting the use of both poison gas and bacteriological methods in warfare. At the 1932–1937 Disarmament Conference, unsuccessful attempts were made to work out an agreement that would prohibit the production and stockpiling of biological and chemical weapons. During World War II, new and more toxic nerve gases were developed, and research and development was begun on biological weapons. The convention was opened for signature at Washington, London, and Moscow on April 10, 1972.

Following the September 2001 events, a Joint Statement by President George W. Bush and President Vladimir Putin on the Cooperation Against Bioterrorism was signed on November 13, 2001: "We agree that, as a key element of our cooperation to counter the threat of terrorist use of biological materials, officials and experts of the United States and Russia will work together on means for countering the threat of bioterrorism, now faced by all nations, and on related health measures, including preventive ones, treatment and possible consequence management".

What Role Can Russia Play in Biological Warfare Nonproliferation and Threat Reduction?

The issues of nonproliferation and threat reduction of biological weapons based on infectious agents are different from other weapons of mass destruction (WMD). These issues are complicated by the general occurrence of pathogens in nature, the continual emergence of new variants of human pathogenic microorganisms and the availability of public health data on the effects of these agents on the human population in different regions of the world. Specific microbial strains and disease incidence data may have the potential for misuse [5].

Russia has conducted significant legislative work to strengthen her compliance with the regime of the 1972 BW Convention, the guidelines of the Australia Group and the BTWC Review Conferences:

- In 1992, a presidential decree was aimed at ensuring the fulfillment of international obligations in the area of biological weapons [6].
- Procedures were introduced for controlling the export from the Russian Federation of disease agents, their genetically altered forms and fragments of genetic material [7] and the relevant amendments were made to the Russian Criminal Code [8].
- Committees on export and currency control, as well as the Committee on the issues of Biological and Chemical Weapons Conventions compliance were formed at the office of the Russian president. Relevant instructions were specified and introduced by the Russian Ministry of Public Health and the State Customs Committee.

Russia has great potential in the area of infectious disease research, as well as in the development and manufacturing of therapeutic and prophylactic preparations at facilities of the Russian Ministry of Public Health, BIOPREPARAT and the local public health establishments. Two large State Research Centers of the Russian Ministry of Public Health – for Applied Microbiology (Obolensk, European region) and Virology and Biotechnology VECTOR (Koltsovo, eastern region) – were involved in biological defense programs of the former Soviet Union before 1990.

The State Research Center of Virology and Biotechnology VECTOR, operated by the Russian Ministry of Public Health, is a large research and production complex, whose primary activities are focused on basic and applied research in the theoretical virology, molecular biology, virology, immunology, aerobiology, epidemiology, and biotechnology. VECTOR also develops and manufactures preventive, therapeutic and diagnostic preparations.

FSU Plans for Future Work Against Bioterrorism

Establishment of a Global Health Disaster Network for natural and man made disasters in the FSU is proposed, as information technology approaches to disaster and bioterrorism mitigation in the FSU barely exist. Given the dearth of approaches in this region, we plan to establish an epidemiologic Global Health Network Disaster program for the Former Soviet Union countries. The fundamental axiom of epidemiology is that adverse health outcomes do not occur randomly within a population but occur in somewhat predictable patterns. Previously, disasters were viewed as "acts of god" or chance occurrence. However, disasters are neither chance nor random events and can be predicted. Once this was recognized, research on the epidemiology of disasters began to flourish [10]. We will establish an active epidemiologic data collection system in the 15 countries in the FSU to collect standardized epidemiologic data and scientifically vet the information from the FSU countries, and distribute it.

This is an inexpensive, low bandwidth approach to collect the data needed at the time of a disaster, and to help officials during and after a disaster mitigate the problems caused by the disaster. Epidemiologic methods can be used to measure and describe the adverse health effects of natural and man-made disasters. Epidemiologic data can be used in designing appropriate warning and evacuation systems, in developing guidelines for preparedness training, and in increasing public awareness through education [10,11]. Our focus is to build the epidemiologic data systems to help mitigate efforts for the famine, earthquake, fire, etc. By building these systems, we will have the added benefit of helping reduce any nuclear or bioterrorism threat escalation associated with an FSU natural disaster. The research will be designed to collect, analyze and distribute epidemiologic data needed at the time of a disaster. The GHDN will provide vital health and mental health data (e.g., potential health risks in disaster-affected countries) and other focused background information in advance of, and just after a disaster to minimize mortality and morbidity and the stabilization of the situation. We will collect information about mental health resources and set the stage for monitoring the psychiatric consequences of disasters. This will require the identification of potentially high risk areas in the FSU for disasters and to target these for contingency planning and advance health information gathering. Such information must be closely correlated with data obtained from the DOD, WHO, UNICEF, Red Cross, and other organizations that track human consequences from disasters. The goal is not to collect new data, but rather, to obtain, synthesize and present data needed for standardized disaster information sharing.

Epidemiologic studies include: surveillance; evaluations of the public health impact of a disaster; evaluations of the natural history of the disaster's acute health effects; analytic studies of risk factors for adverse health effects; clinical investigations of the efficacy and effectiveness of particular approaches to diagnosis and treatment; population-based studies of long-term health effects; studies of the psychosocial impact of a disaster including fear management; and evaluations of the effectiveness of various types of acute care and the long-term effects of disaster relief aid.

Specific aims may include:

- To establish an expert task force of global leaders including those in psychiatric epidemiology to map scenarios, and then identify potential epidemiologic data which would help to prevent and/or mitigate these events.
- To identify standardized epidemiologic data and metrics to evaluate usefulness, accuracy and cost, including mental health epidemiology.
- To build a Former Soviet Union (FSU) Global Disaster Network on the Internet.
- To monitor the potential and early impact of disasters for better preparation and response. Build Just-in-Time lectures providing background information for the disaster foci area, the country and globally in both English and the Native language, the data will be derived from other agencies (e.g. Relief Web) or government organizations.
- To evaluate the system.

We propose to build a network of scientists engaged in disaster epidemiology in a manner similar to the parent Supercourse. First, we will identify those individuals involved in areas related to disaster epidemiology who are already members of the parent Supercourse, and link this network in the FSU with those in the US and elsewhere. Moreover, this group will recruit their friends and colleagues who are engaged in disaster epidemiology. A network such as this with cross-country communication is important to disaster mitigation and for keeping a dialogue open under adverse political times as well. We already have people trained in epidemiology and public health in all 15 countries. These individuals will serve as our "ground truth" in each of the countries. They will collect data from the government, academia and other places concerning disasters within their countries. The data will be updated annually and shared. Data to be included:

- Epidemic Risks.
- Incidence of communicable diseases and vaccination coverage.
- Nutritional status.
- Country health profile.
- Basic ethnographic data on populations at risk of disasters.
- Database of in-country NGOs and UN agencies and a catalogue of their resources. Ideally this would include key individuals and points of contact.
- Description of that countries disaster plan (if any).
- Organization of health professional training in the country (e.g. amount of training A1, A2, A3 level nurses in Rwanda, how many years of university education and medical school a clinical office in Kenya has, and the type of post-graduate data about health professional education that would help in establishing training programs f or health workers that are compatible with training in the US).
- Logistics: warehouse capacity, availability of gas/diesel, air/road access, telecommunications, etc. Again, while gathering information for a whole country might not be possible, it might be feasible for border regions.
- Local and regional laboratory capabilities to identify the causative agent.
- In-house production capability for availability of drugs, jerry cans, cooking kits, etc.
- In-country availability of food stocks.
- Data concerning political stability.
- Epidemiologic data on nuclear capacity.

It is impossible to predict where a disaster will occur and the exact data available and needed. We expand the network of people skilled in epidemiology across the FSU. Should a disaster occur, this network will be mobilized. Thus, should there be another earthquake in Armenia, we will be able to mobilize 20 people in Armenia who have epidemiologic training who can help collect data, have contacts in other places in the FSU, and who are networked to the coordinating center in Pittsburgh. Data will be collected by these individuals, and data analyzed within the country as soon as possible. Should there be problems in the analysis, the experts in Pittsburgh and the task force will assist.

The concept of Just-in-Time lectures comes from manufacturing. We have identified over 10,000 people world wide who are our suppliers of lectures. Should an Anthrax attack occur, we can immediately identify the experts and work with them to construct background lectures, for example, on the epidemiology of anthrax, as well as lectures that describe the event as it is unfolding. Thus, we will cobble together information gleaned from the scientific journals, as well as to obtain "on the ground reporting" which represents health officials on the ground in order to rapidly put up a lecture and to continuously update the lecture until the events are resolved. We will obtain information from our network and CDC, FEMA, ARC, who are at ground zero.

These lectures are about a single acute event, such as Chi-Chi's, or Anthrax. With the JIT lectures we need to assemble to the knowledgeable scientists from our network and speed the processing of the lectures to maximize the rapid distribution of the rapidly decaying lectures. We envision 4 Supercourse JIT/CNN lectures on these topics a year from the FSU.

Prevention of Terrorism

Because of its uncertainty, it is hard to collect unbiased data on terrorism and it is hard to propose specific prevention measures. Below, we present the general approaches for prevention from the epidemiological and public health perspective.

Primary prevention

- Education!!!
- Understand differences in cultures, religions, beliefs and human behaviors.
- Think of peace, freedom and equality of all human beings, not just "my group of people."
- Eliminate the root causes of terrorism.

Secondary prevention

- Establish surveillance and monitoring systems on terrorism attack.
- Improve protective systems for citizens.

Tertiary prevention

- Early detection of the sources.
- Prevent the extension of impairments.
- Rescue the survivors.
- Console the rest of the population.

Conclusions

Terrorism is an unlawful act with a long history of being used to achieve political, religious and ideological objectives. It can be conducted through firearms, explosive devices, and biological, chemical, and nuclear agents. Despite the events of 2001, the risk of dying from terrorism has remained much lower than that from motor vehicles, smoking, and alcohol use [9].

References

[1] J. David Piposzar. Public Health Response to Terrorism – Preparedness. www.pitt.edu/~super1/lecture/lec0911/index.htm.
[2] Fred T. Muwanga. The History of Bioterrorism. www.pitt.edu/~super1/lecture/lec11591/index.htm.
[3] Rashid A. Chotani. Identification of Bioterrorism Agents. www.pitt.edu/~super1/lecture/lec10181/index.htm.
[4] Phillip L. Coule. Chemical and Biological Terrorism, an overview of the threat. http://www.pemba.utk.edu/bt101/Chem_and_Biol_Terrorism-_Dr._Coule.ppt.
[5] Sergey V. Netesov, Lev S. Sandakhchiev, State Research Center of Virology and Biotechnology VECTOR, Russian Ministry of Public Health, Koltsovo, Novosibirsk Region, Russia. The Development Of A Network Of International Centers To Combat Infectious Diseases And Bioterrorism Threats Applied Science and Analysis, Inc. The ASA Newsletter.
[6] Decree of the Russian President on ensuring the fulfillment of international obligations in the area of biological weapons, Decree 1 390, April 11, 1992.
[7] Procedures for controlling the export from Russian Federation of disease agents, their genetically altered forms, and fragments of genetic material that can be used for developing bacteriological (biological) and toxin weapons, Nov. 20, 1992, Decision RG 1 892.
[8] Penalties for crimes against peace and security of mankind: Production or proliferation of weapon of mass destruction. Sections 355 and 356 of the Russian Criminal Code. 1996.

[9] Supercourse Team. Understanding September 11. http://www.pitt.edu/~super1/lecture/lec6651/001.htm.
[10] Noji E.K. Disaster epidemiology: Challenges for public health action. J Public Health Policy. 1992; 13:332–240.
[11] Binder S., Sanderson L.M. The role of the epidemiologist in natural disasters. Ann Emerg Med. 1987; 16:1081–4.

5. Modeling and Simulation

Strengthening National Public Health Preparedness and Response to Chemical, Biological and Radiological Agent Threats. Edited by C.E. Cummings and E. Stikova.
IOS Press, 2007.

Using Modeling and Simulation in Planning for Public Health Emergencies

C. Donald ROBINSON
Department of Systems and Information Engineering
University of Virginia, Charlottesville, VA, USA

Abstract. Modeling is generally used to study complex systems. Furthermore, computer simulation of models is a popular way to examine and study analytically intractable models. While these practices are quite popular, it is important to use them judiciously. Models can be created that cannot be verified or validated. Furthermore, using simulation results without proper analysis can lead to false claims. Also, an emerging field of modeling, agent-based modeling, is a field that deserves special attention due to the complexity of the behaviors modeled. Agent-based models define rules for agents to follow and their interactions are then simulated to observe the effects of their resultant behavior. An agent-based model of epidemic spread through a population is presented which leads to some unintuitive results regarding the effects of immunization.

1. Using Simulations in Modeling Complex Systems

The threat of public health emergencies is certainly increasing. With terrorism, increasing population densities and new pathogens emerging, proper public health response to emergencies has never been needed more. However, what is a proper or even optimal public health response? Another issue is that even if an optimal response existed, how does one carry out the plan? True emergencies are no place to test out ideas for response. During a real emergency public health officials and responders would hopefully only be implementing planned procedures. Modeling and simulation should serve as a natural means of testing and verifying potential procedures to be used in a real emergency.

Modeling and simulation provide a means of exploring the possibilities of public health response as well as the potential effects of an emergency in a safe, fast, and relatively cheap method. Live simulations are events in which real responders train through a mock scenario where most variables are carefully controlled. However, live simulations take exorbitant amounts of time and money to implement. Furthermore, a live simulation exercise gives a researcher only one data point. Granted many variables can be measured in a live simulation. Nonetheless it is one response to one scenario. While the training received by responders in a live scenario might be invaluable, the benefit to researchers is less so.

1.1. Modeling

What is modeling? Quite simply, modeling is a simplification. A model airplane is a simplification of a larger more complex system. A computer game is a simplification of a potentially more elaborate real world scenario. Furthermore, a mathematical equation might be a simplification of a much more complex process. For years it was though that momentum was equal to the mass of an object multiplied by its velocity or rather $p = m\mathbf{v}$ where p is the momentum of an object, m is the mass and $\mathbf{v}$ is the velocity of the object. This paradigm of science was thought to be law. However, Einstein's theories of relativity showed that it was a simplification of a more complex process. The relativistic calculation includes the parameter c which is the speed of light. The equation of momentum is

$$p = \gamma m\mathbf{v} = \frac{1}{\sqrt{1 - v^2/c^2}} m\mathbf{v}$$

Thus the Newtonian equations of momentum are simply a model which is valid assuming the relative velocity of the object is substantially less than the speed of light. However, when the velocity approaches the speed of light, the extra term in the relativistic equation alters the momentum significantly.

It is necessary to note that because a model is a simplification of a more complex process, the model will not always be right. George Box once said "All models are wrong, but some are useful" [1]. What this means is that a model and simulation of an epidemic spread through a city will not necessarily yield an exact prediction of the true underlying process. Instead a model could yield a distribution of possibilities of an epidemic spread. One might spend millions of dollars to create a simulation of a public health emergency. It might be tempting to run the simulation, look at the output and declare that the output was in fact what a responder could expect. This is certainly not the correct method of using a simulation. One must test the validity of the assumptions made in the models, explore the possible outcomes over the entire range of desired possible inputs, and determine a distribution of possible outcomes.

1.2. Agent-Based Modeling

One modeling tool that is being used extensively is that of agent based modeling [2]. In agent based modeling the researcher models the individual behavior or a set of agents and essentially lets them go in a simulated environment and observes the resultant behavior. An example to this would be simulating individual birds in a flock. One approach is to just model and simulate the movement of the flock as if it were a single entity. Characteristics might include the velocity and size of the flock. This is clearly a simplification of the real processes going on which create a flock. An agent-based model would simulate each bird in the flock. The model would include rules for following or leading in flight, rules for determining individual speed and spacing, as well as rules which would govern whether a bird would stay in the flock or branch off. Then the modeler would simulate a bunch of these agents interacting together. The result of such a simulation would be the very characteristics of the simpler model mentioned before. These output characteristics would be the general velocity and size of the flock. However, these outputs are the result of a more individualized modeling effort.

Some might say that agent-based models are more complex than more aggregated models. While this is sometimes the case, it is certainly not generally true. A good example is John Conway's "Game of Life"[3]. This model is a model of a population movement. It is set on an infinite discrete matrix with each cell either being alive or dead. The model has four simple rules. Any live cell with fewer than two neighbors dies of loneliness. Any live cell with more than three neighbors dies of overcrowding. Any dead cell with three neighbors comes to life. Lastly, any live cell with two or three neighbors lives. One can find simple initializations of this model and simulate the resultant behavior. More often than not, the result is a highly dynamic and complex behavior of the population. An example of a very dynamic and sustaining behavior is that of the Gosper Glider Gun. In this example, the result is a complex pattern of masses converging and "shooting off" of smaller populations that continue forever. Thus, agent-based modeling can either involve very simple or very complex models. However, the results, when simulated can be very complex and unexpected.

1.3. Which Simulations to Run

Simulations are often thought of as black boxes. A set of inputs is fed into this black box and outputs emerge. The calculations that take place within the simulation are the model of the underlying system under study. Inputs might be factors like the type of epidemic spreading through a population, the number of responders able to assist, the capabilities of the responders, or even the parameters governing how the pathogen might spread from one person to another. Outputs of the simulation might be the total number of people infected, the mortality rate, or the time until the epidemic is stopped.

This framework of a simulation is very useful. What is generally desired in simulations is either to explore a parameter space or to find an optimal solution. The aforementioned inputs to a simulation define the parameter space. This is the set of all possible ways to initialize a simulation. Exploring a parameter space involves running the simulation with many different settings in this space and observing the outcome. Often a researcher will take the outcomes and fit another model to them. An example of this might occur if a simulation takes a long time to run. In this case a researcher might try to fit a meta-model to the simulation outputs and use this as a surrogate for the simulation itself [2]. A closed form solution is nearly instantaneous to calculate. Thus, this surrogate model would be used in place of the simulation. Keep in mind though, that the simulation is used in place of reality. Thus, in this case the simulation is a model of reality, and the surrogate is a model of a model. A simplification of a simplification will probably yield more error in the results and this should be taken into account.

1.4. Experimental Design

Exploring the parameter space normally involves a design of experiments. Many different types of experimental designs exist and this is its own field of research. There are many different criteria for defining an experimental design. Some seek to test the extrema of the input space in order to estimate overall effects [4]. Other experimental designs are aimed to sample uniformly from the parameter space. These designs are most commonly referred to as space-filling designs [5]. Whatever the criteria, it is important for a researcher to make the most judicious use of the runs executed on a simulation. Simulations necessarily take time to compute. Some complex simulations can take hours, days, or even weeks to calculate. Furthermore, a researcher necessarily has some sort of budget or constraint with which to simulate. This constraint might be time or might be money. Nevertheless, this constraint always exists and necessitates the use of efficient experimental designs. Some prominent examples of experimental designs include Latin hypercube designs, fractional factorial designs, and design with so-called alphabetic optimality [6]. Each of these has a different mechanism for determining which sets of inputs, or sample points, a researcher will give the simulation to calculate a value.

Designs that make the most efficient use of these sample points are extremely important. One reason is due to the curse of dimensionality. The curse of dimensionality can be summed up as saying that as the number of dimensions in a problem increases, the number of sample points needed to "describe" the outputs given a parameter space increase exponentially. Consider an example that will use a full factorial design. Given the parameter space is defined as a hyper-rectangle (high dimensional cube), factorial designs essentially test just the corners of this hyper-rectangle. One of the reasons for this is to fit a simple linear model to the responses and determine which inputs are most significant in the problem. Now also consider that the simulation in question takes about ten seconds to run one sample point and get a response. If the problem is one-dimensional, meaning there is only one input such as the population in a city then a factorial design will test two points and take 20 seconds. If the problem is two dimensional, the inputs might be the population of a city and the probability of spreading the disease. In this case a factorial design will test four points and take 40 seconds. In a 10-dimensinoal problem the process would test 1024 points and take about 2 hours, 50 minutes and 40 seconds. A 100 dimensional problem would take about 4×10^{23} seconds. This is older than the estimated age of the universe. It is easy to imagine 100 potential inputs that would affect the spread of an epidemic through a population. In fact models have been made that include thousands of variables. While these models might be very accurate, it is difficult to test the models and to quantitatively verify whether those 1000 variables do in fact represent the process being modeled. Thus, Occam's razor should apply and the most parsimonious model should be considered the best.

1.5. Optimization

Optimization is another primary objective of modeling and simulation. Optimization implies searching the parameter space to find the input or set of inputs which yield an output that has a maximum (or minimum) value of some associated measure [4]. This measure could be that actual value of the output or could be another measure such as the variance associated with the output. For example, Taguchi designs are based on maximizing a response while also finding a solution where a slight change in the input parameters will still yield a robust maximum [7]. In other words, if the parameters move slightly away from the optimal, the response doesn't plummet off of a cliff.

Optimization is often very tricky in simulations. The common problem of local versus global optima is a problem. Local optima are maximums only for a small area or subset of the parameter space. Global optima are maximized for the entire parameter space. However, other problems of optimization also exist. One such problem is that of determining what to optimize. Clearly in a public health emergency there are many criteria that need to be optimized simultaneously. Examples include the speed of response, minimizing the number of people infected or exposed, minimizing mortality rates, and maximizing public relations sentiments. The field of multi-objective optimization has many methods to deal with these types of problems. Nevertheless the assumption is that a measure can be calculated for each output that can be optimized and a method is used to do it.

2. Agent-Based Simulation of Epidemic Spread

The author has created an agent-based model of epidemic spread that did in fact demonstrate some interesting results. The model has four basic parameters. The spread of the disease is modeled using the following process. A population moves throughout a city randomly. The movement is modeled by a simple Gaussian random variable. In each discrete time step, every agent moves according to a random draw from a non-correlated bivariate Gaussian distribution with the mean equal to the current position. After each time step all of the agents interact. If a person is within a specified distance of another person, then it is possible for the disease to be transmitted if it is present between any one of two agents. Thus, if agents are not within this distance of one another, there is a zero probability of the disease propagating. Next assume two agents are within this transmittal distance and one has the disease and the other does not. There is a probability that the infected person will transmit the disease. This probability could be modeled as a function of the incubation time of the disease or other parameters. However, in this model it is constant. Next, if the disease is in fact transmitted to the other person then there is another probability that the uninfected person receives and contracts the disease. Again this could be the function of the person's resistance to the disease or other factors but for this example it is constant. It is lastly assumed that once a person contracts the disease, there is no cure and the person is infected for the duration of the simulation. Furthermore, each simulation begins with only one person infected. Thus, for every test, the simulation begins with one person infected and the spread is observed from that initial person transmitting the disease.

Lastly, there is a parameter that determines what percentage of the population is immunized. It is assumed that if a person is immunized that it is impossible for he or she to contract the disease, regardless of the proximity to another infected person. The parameter governing this is a random proportion of the population that is immunized. Thus, when the simulation is initialized, a random proportion of the agents are selected independently and immunized. Then the simulation is set in motion.

This gives us the four parameters of the simulation: distance of transmittal, probability of transmittal, probability of reception, and proportion of the population that is immunized. The environment of the simulation is based on data layers from a geographic information system (GIS). The setting is actually a 121 square mile area of Northridge, CA and consists of 16767 agents in this area. The actual population of this area in 2000 was approximately 100,000 people according to the U.S. Census. The distribution of these agents in the Northridge area is determined according to Census data. Each census tract has approximately the right proportion of the agents within it to mimic a full population. Thus, if one census tract contains 10% of the total population of the Northridge area, then that census tract will be initialized with approximately 1677 agents or 10% of 16767. The simulation then progresses through 200 time steps and the spread of the disease is observed.

In testing the simulation 10 random experiments were performed each with a different set of parameters. Each parameter was given either 2 or 3 levels on which to test. The distance of transmittal was set at either 100 or 500 meters. It is acknowledged that this distance is unlikely in a real epidemic. However, the population used was approximately 17% of the total population so larger distances were used to model the spread with a decreased population. The population size was only 17% of the true population for computational efficiency. The probability of transmittal was set at either 10% or 80%. Thus, if the probability of transmittal is 80%, then there is a random long-run probability of 80% that the infected person will transmit the disease to an uninfected person within the threshold distance of transmission. The probability of reception was set to either 10% or 80% as well. Lastly, the percentage immunized was either 0%, 30%, or 75%.

To denote each experiment the following nomenclature is used. The letter 'T' will represent an 80% probability of transmittal and the letter 't' will represent a 10% probability of transmittal. Similarly, 'R' represents an 80% probability of reception and 'r' represents a 10% probability of reception. The letter 'I' represents 75% of the population being immunized. Also, 'iI' will represent 30% and 'i' will represent 0% immunizations. Lastly 'D' will represent a 500 meter buffer for transmittal and 'd' will represent a 100 meter buffer of transmittal. Thus, the experiments performed are as follows:

t.r.i.d	
T.R.i.d.	t.r.iI.d.
t.R.i.d.	t.r.iI.D.
T.r.i.d.	T.R.iI.d.
t.R.I.d.	T.R.iI.D
T.r.I.d.	

Table 1

One can see that the first 4 experiments are testing the differences in the transmittal and reception frequencies. The fifth and sixth experiments are meant to observe the differences in high transmission versus high reception with an immunized population. Lastly, the last four experiments are meant to observe the effects of distance on high and low transmittal and reception rates. Some interesting results emerge from these few experiments.

To show some of the results of the disease spread I will primarily look at two metrics. The first is the total number of infected people at each of the 200 time steps. The other metric is essentially the differential of this metric. Thus, the second metric is the number of people who became infected in each time period. Plots of these two time series for the T.R.I.d. experiment are shown in figures 1 and 2.

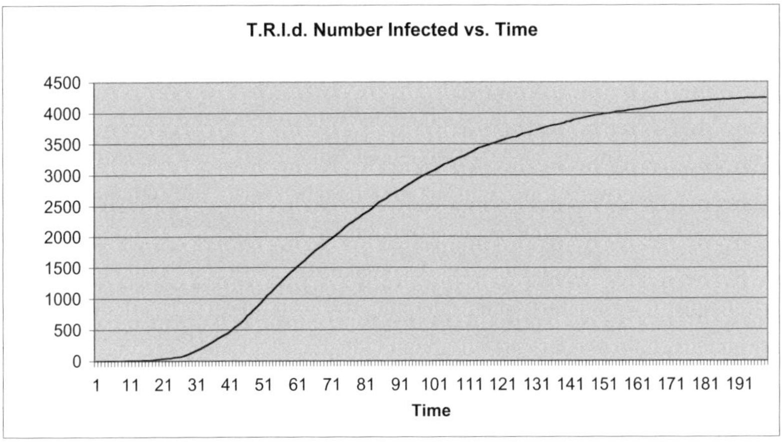

Figure 1

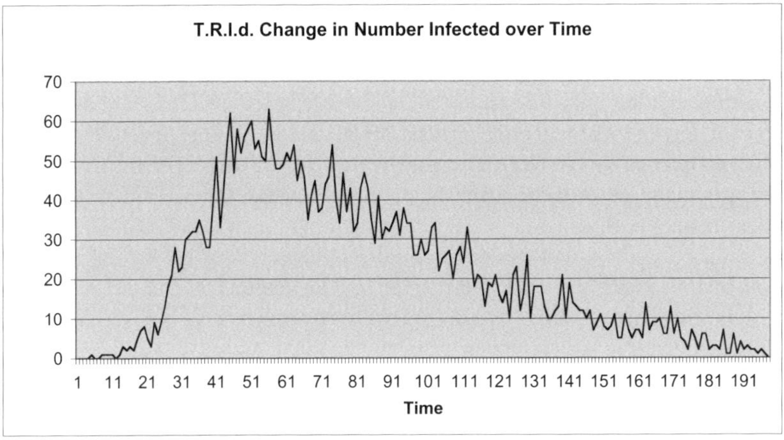

Figure 2

The same plots for the t.r.i.d. experiment are shown in figures 3 and 4.

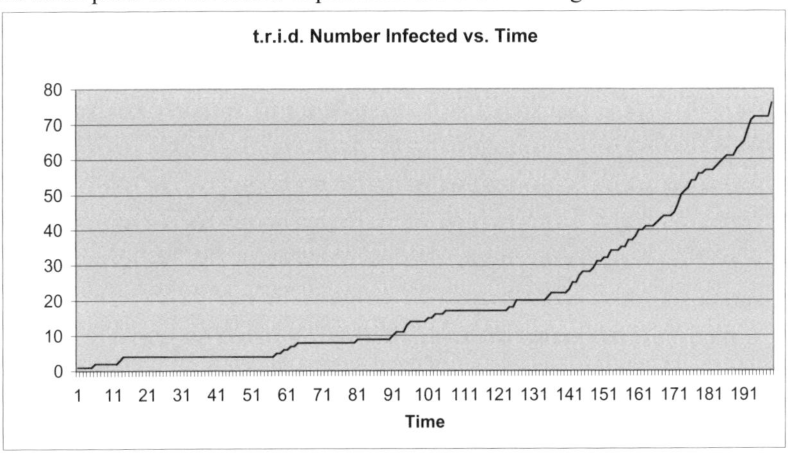

Figure 3

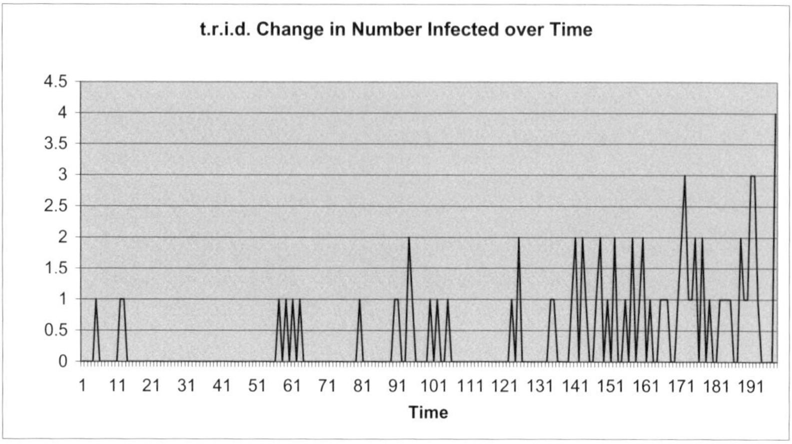

Figure 4

Here one can see that the scales are quite different. High transmittal and reception rates, even with high immunization rates yielded a total of 4243 out of 16767 people infected in 200 time steps. However, low transmittal and reception rates even with no immunization yielded only 76 out of 16767 people infected. Thus the differential plot for the t.r.i.d. experiment has only four possible values. This means that at a maximum only four people contracted the disease in any single time period. Whereas in the T.R.I.d experiment, 60 or more people contracted the disease in a single time period on three occasions.

It should be noticed that the differential plots there is usually a hump in the plots. The peak of this hump represents the time period where the disease is spreading the fastest. For example in the t.R.i.d experiment versus the T.r.i.d experiment, the "humps" are located in different time periods.

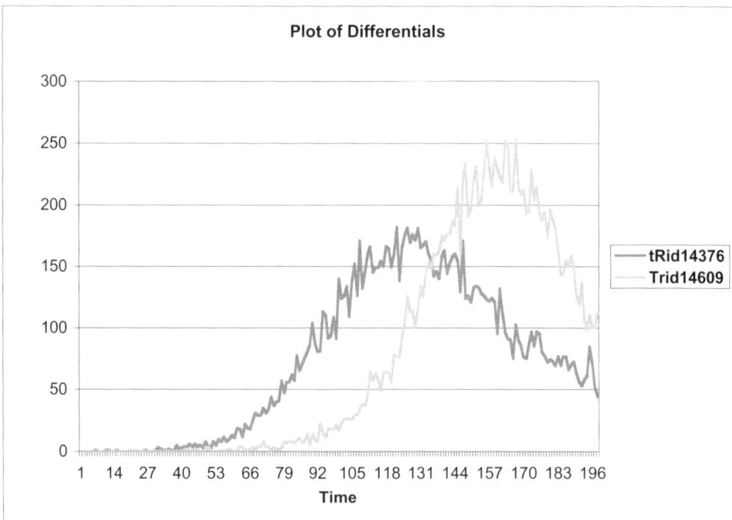

Figure 5

In figure 5, the differential plots for these two experiments are shown. Note that the legend in the figure shows each respective plot with the total number infected. Thus, the t.R.i.d. experiment yielded 14376 total infected whereas the T.r.i.d. experiment yielded 14609 total infected. These numbers are very close to each other showing that in 200 time steps the overall number of infected is about the same. However, it is seen that a high probability of reception results in more people being infected sooner. Rather a high probability of transmittal results in a more rapid rate of people infected after a longer period of time. In other words, if the disease doesn't transmit well but the population is very susceptible if it is transmitted, then there will be a gradual increase in the number of infected people seen during any time period. However, if the transmission rate is high but the population is fairly resistant, and if nothing is done to stop the disease, there will be a rapid increase in the numbers infected and there will be a large spike in the numbers infected. This is seen around the 160th time period in figure 5. This has implications for what hospitals might see and can help prepare them for what to see for certain types of diseases.

One last interesting point that resulted from this model deals with immunization rates and the speed at which the disease spreads through the population. It is assumed in the model that if a person is immunized then he or she will not be able to become infected. Thus, if 75% of the population is immunized, then only 25% of the remaining population is susceptible to the disease. In looking at the differences in immunization rates only the percentage of the susceptible population is measured. Figure 6 shows the percentage of susceptible people infected given a low transmission rate and a high reception rate. The immunization levels are 0%, 30%, and 75%.

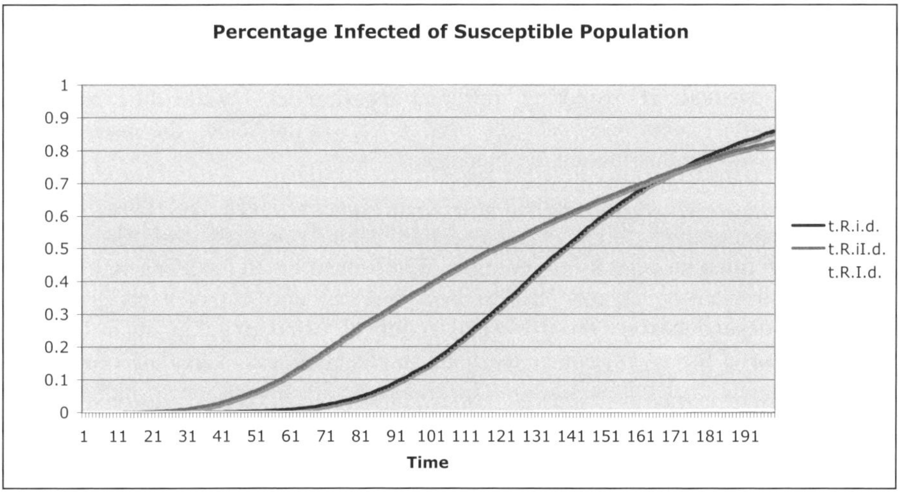

Figure 6

What is interesting is the difference between the t.R.i.d. experiment and the t.R.iI.d. experiment. It is logical to think that immunizing people would make the disease spread more slowly through the susceptible population. If fewer people are susceptible to contracting the disease then there will be fewer people to spread it which should mean a slower spread throughout the population. However, the two aforementioned experiments first and foremost led to about the same percentage of susceptible people being infected after 200 time steps. What is interesting, though, is that the disease seemed to spread faster through the susceptible population with 30% immunized versus having no one immunized. This can be seen in that the line in figure 6 depicting the t.R.iI.d. experiment is above the t.R.i.d. line for a good portion of the total time periods. In fact at time period 100, 14% of the susceptible population with no one immunized had contracted the disease. However, 39% of the susceptible population had contracted the disease with 30% immunized. Thus, agent-based modeling can sometimes go against intuition and yield interesting results. The interesting result here is that by immunizing 30% of the population versus 0%, fewer total people contracted the disease, but the disease spread faster through the susceptible population.

It should be noted that these sets of experiments on this model do not conclusively validate some of the findings in generality. Rather this is just an example of possible effects that are not necessarily intuitive. Future work would include testing many more points and furthermore replicating experiments at each set of inputs. The replication is necessary due to the stochastic nature of the simulation. While this example is certainly not sufficient to make real decisions in the wake of a spreading epidemic, it does provide very useful insight into directions for future research.

3. Conclusion

Overall, any modeling technique in planning for public health emergencies should be used with caution. Every model is a simplification of reality and thus cannot predict perfectly what the outcome of any situation will be. However, modeling efforts can help researchers and practitioners alike understand the possible outcomes and the general implications of different ideas, methods, diseases, etc. Simulation is an additional tool that many can use. Simulation provides a relatively cheap and noninvasive method of exploring different possibilities. While like modeling, simulation results cannot necessarily be used as perfect predictors, the results can be used as a guide to the distribution of outcomes.

Lastly, agent based modeling is a method in which the rules governing behavior often intuitively follows the way certain natural processes work. However, the outcomes can often be contrary to intuition. While agent based modeling is sometimes difficult to implement, it can lead to new insights into complex processes that particularly depend heavily on individual action or behavior rather than aggregate tendencies. All of these tools can be used and should be used to help plan for emerging threats in the field of public health. The public health area is one which has dire consequences if handled improperly and one must use all tools available and within budgetary constraints to achieve as close to an optimal solution as is possible.

References

[1] Box, George. 1979. Robustness in the strategy of scientific model building. In R. Launer & G. Wilkinson (Eds.) *Robustness in Statistics,* cited by Temple, J. (1998) Robustness tests of the augmented Solow model, *Journal of Applied Econometrics*, 361-75.

[2] Sanchez, Susan and Thomas Lucas. 2002. Exploring the World of Agent-Based Simulations. *Proceedings of the 2002 Winter Simulation Conference*, San Diego, CA.

[3] Axelrod, Robert. 1997. Advancing the Art of Simulation in the Social Sciences, In *Simulating Social Phenomena*, ed. R. Conte, R. Hegselmann, and P. Terna. Springer-Verlag:Berlin.

[4] Myers, Raymond H. and Douglas C. Montgomery. 2002. *Response Surface Methodology*, Wiley and Sons: New York.
[5] Santner, Thomas J., Brian J. Williams and William I. Notz. 2003. *Design and Analysis of Computer Experiments*, Springer: New York.
[6] Atkinson, Anthony C. and Alexander N. Donev. 1992. Optimum Experimental Designs, Vol 8 of Oxford Statistical Science Series. Oxford University Press: Oxford.
[7] Park, Sung H. 1996. *Robust Design and Analysis for Quality Engineering*. Chapman and Hall: New York.

Strengthening National Public Health Preparedness and Response to Chemical, Biological and Radiological Agent Threats. Edited by C.E. Cummings and E. Stikova.
IOS Press, 2007.

Modeling Munitions and Explosives of Concern (MEC) CBRN Hazards: Novel Tools and Approaches for Strengthening the Conceptual Site Model for Public Health Preparedness

Chuck TOMLJANOVIC, MSEM[a,b] and Conrad VOLZ, DrPH, MPH[b]
[a]*Concurrent Technologies Corporation, 100 CTC Drive, Johnstown, PA 15904*
[b]*University of Pittsburgh, Pittsburgh, PA 15261*

Abstract. Munitions and Explosives of Concern (MEC), including CBRN Unexploded Ordnance (UXO), pose a significant health hazard to human health and the environment. Contamination from training, violent war fighting, and emplacement activities is world-wide and estimated in millions of acres. Mitigating the physical and chemical hazards of MEC within NATO and its Partnering Countries, including areas within the Balkan Peninsula, will be extensive and cost tens of billions of dollars. Conceptual site modeling (CSM) within the environmental community has been a tool for environmental remediation specialists to help ascertain contaminant footprints, determine most likely transport and exposure pathways, identify exposed populations (human and ecological), and focus remediation/restoration efforts. These same CSM principles have been applied to UXO remediation efforts. However, recently, new tools have been developed by the United States (U.S.) Department of Defense (DoD) through the U.S. Army Corps of Engineers (U.S. ACE) to support data collection and management of MEC response at Munition Recovery Sites (MRS). Novel application of these tools, either alone or in tandem, can be used to develop advanced and strengthened CSMs. Application of these tools could play a key role in better understanding the true nature of MEC/UXO, thereby improving response efforts at MRS by NATO and Partnering Countries. This paper explores the potential application of these tools specifically in strengthening the CSM for munition response, and proposes approaches that could be potentially used to improve the public health preparedness and efficient response to CRBN MEC within the Balkan Peninsula.

Keywords. Munitions, UXO, MEC, Remnants of War, Conceptual Modeling

Background and Problem Statement

Munitions and Explosives of Concern (MEC), including Unexploded Ordnance (UXO), emplaced landmines, and other remnants of war (ROW), is a costly problem to the United States (U.S.) Department of Defense (DoD), as well as to NATO Partnering Countries (NPC). Generally, current impact areas are often a result of past live fire activities and/or emplacement of ordnance devices. The MEC problem itself is difficult to manage because contamination is so extensive world-wide and total MEC/ROW tonnage is unknown. Moreover, because MEC and ROW contamination is hazardous to public health and the environment (e.g., direct physical hazards/human "explosion risk" and "eco-hazardous munitions risk"), it has become an environmental response priority to many including world-wide military and humanitarian organizations.

MEC is very costly in terms of maintaining active training ranges for the war fighter, realigning former military lands undergoing reclamation, and minimizing public health injuries. The Final Report (2003) of the Defense Science Board Task Force on UXO expands on the massive scale of the problem within the U.S., which involves over 10 million acres of land from over 1,400 UXO sites in the United States alone. They report that current estimates to effectively respond to the UXO problem inside the Continental United States (CONUS) is tens of billions of dollars ([1]).

The basis of this presentation and paper is the hypothesis that improved environmental conceptual modeling will play an increasing and important role in improving detecting, characterizing, remediation, and managing UXO response efforts (especially in the Balkan regions where little or no response structure exists) there by reducing the resource requirements. NATO Partners must work to assure that MEC/UXO recovery efforts and technologies selected and implemented across the world are effective in both saving time and money to allow the redirection of any savings to meet other, high-priority humanitarian needs. Through the conceptual modeling of UXO CBRN hazards (e.g., UXO recovery depths and recovery densities), along with other relevant site information, environmental specialists can establish a standardized foundation across multiple sites in collecting UXO recovery data to better understand and to respond to the UXO hazard as well as for management processes at varied MEC/ROW response efforts in protecting public health.

Extent of Contamination in NATO Areas of Concern — The Balkan Peninsula

Contaminated areas of particular interest to this paper and presentation are select locations within Albania, Bulgaria, Montenegro, Macedonia, Turkey, Slovenia, Boznia/Herzegovenia (BiH), Croatia, Greece, Serbia, and Romania (Fig. 1). Many of these countries have disproportionate resources available for munitions response compared to the extensive contamination, and as such, have limited compilations of the extent and type of MEC contamination. Nonetheless, some response agencies within countries have implemented response actions and worked with varied agencies within the U.S. and abroad to remediate directly affected areas and secure funding for response needs.

Specifically, Albania, BiH, and Kosovo, Serbia are three countries from the Balkan Region that demonstrate response efforts where some data exists signifying the extent of the MEC/ROW problem in the Balkans. According to the United Nations Mine Action Service and Landmine Monitor ([3,4]), the Albania Mine Action Programme (AMAP) is working toward remediating up to 15.3 million square meters (M^2) of impacted areas in Albania that adversely affects almost 40 villages.

Since 1999, over 30 people have been killed and 238 injured of the 25,500 directly affected individuals in Albania. In 2006, $7.98 million dollars (U.S.) was available for response needs for the AMAP. In BiH, the Ministry for Civil Affairs is remediating 2.1 Million M^2 of wide-spread MEC/ROW contaminated land that impacts 1,366 communities. According to the same sources, in BiH, There have been over 150 related MEC/ROW incidents since 2004 ([5,6]). In 2006 $6.69 M (U.S.) dollars were available for response needs in BiH. Finally, in Kosovo the Kosovo Protection Corps. (KPC)

Figure 1. Past violent activities have lead to wide-spread MEC/ROW contamination in the Balkan region. This contamination poses significant hazards to human health and the environment ([2]).

have cleared approximately 2 million M2of impacted area through 2005. Even with these response actions, over 40 dangerous areas remain. Most recently, there have been 11 incidents in Kosovo with 13 injured and 1 fatality. In 2006, $0.54 M (U.S.) dollars were available to the KPC for MEC/ROW response needs.

Because each site holds hidden dangers and uncertainties (e.g., depths of ordnance pieces underground are generally unpredictable), munitions response in these countries is time consuming and becomes very costly in order to be effective. Improved methods for response to the human and ecological hazards from MEC are needed to prioritize actions and improve efficiencies in order to allow the maximum response possible to this extensive problem. Improved conceptual site modeling (CSM) offers responders the opportunity to conceptually assess all the possible exposure hazards and risks from MEC/ROW, including the direct physical hazards from explosions and eco-hazards from munition infiltration to the surrounding environment (i.e., soils and groundwater). An improved CSM that presents the true nature of UXO for a specific site can become the focal point for any MEC/ROW site assessment and related response activities.

Conceptual Modeling for Environmental Response

Conceptual modeling has become a powerful tool for remediation in the environmental arena. Using conceptual models, environmental scientists and public health professionals can conduct important analyses to identify sources of contamination and direct physical and chemical public health hazards. Also, health scientists can identify contaminant transport and fate mechanisms, list pathways and routes of exposure or potential chemical uptake and, suggest potential underlying patterns and trends in order to support environmental decision making. Modeling contaminants and pathways along with other important site attributes allows decision makers to identify the underlying challenges at hand from the particular environmental setting and help improve decision making as well as the communication of risk to parties involved.

Widely-used, generally-accepted guidance has been developed for developing CSMs and is referenced throughout literature ([7–10]). Although this is the case, CSMs vary greatly in format, style, and content (Fig. 2). They vary greatly because they are

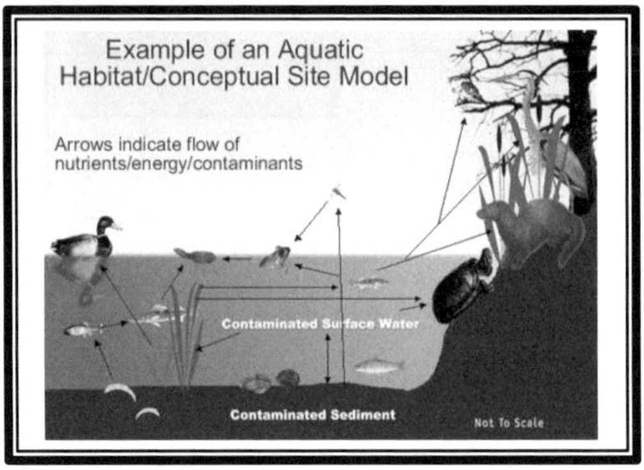

Figure 2. Example Conceptual Site Model (CSM) for depicting site conditions. CSMs depict the interaction of the contaminant of interest with the human and ecological environment. CSMs often differ in format, style, and content ([8,9]).

highly dependant on the level of site understanding and site assessment/project need. Moreover, CSMs must also remain dynamic throughout the life-cycle of an environmental response project – as more and more information becomes available, the CSM is continuously improved Not only do CSMs describe what specialists know about a site, some of the advanced CSMs also describe the uncertainty and limitations associated with site activities and response. And although they all do differ to some extent, all CSMs generally present a primary source, basic release mechanisms, transport pathways and end-receptors. Moreover, they are all used to support effective site management and risk response.

Managers from MEC/ROW sites also leverage conceptual modeling procedures and processes in place for munition response sites (MRS) and associated response activities (Fig. 3) ([11,12]). Although similar in basic structure and function to a traditional CSM found at a chemical response site, they hold a primary difference. That difference is in presenting information on receptor interaction with site MEC/ROW hazards. At a MEC/ROW site conceptual models specifically call out *hazard access* (i.e., noting if access to the ordnance pieces is available), and identifying related *site activity* (i.e., noting if the activity leading to exposure is invasive or non-invasive). (These data points are closest related to the release mechanism under the traditional CSM for a modeling and assessing a chemically contaminated response site.) It is an improved MRS CSM that holds the greatest potential for assuring response activities are the most efficient during response planning and action.

Improved Conceptual Modeling for Munitions Response

The physical setting and release profiles of ordnance exposures within the MRS CSM can be further refined to improve MEC/ROW response actions. However, this will require the efficient and effective collection and sharing of MEC/ROW recovery data points for presentation within the CSM. World-wide, MECs and ROW are recovered in various forms and depths. A central data housing unit promoting consistency in data collection, management, and analysis within the Balkan Region can dramatically improve CSM development and thus response at MRS in this area.

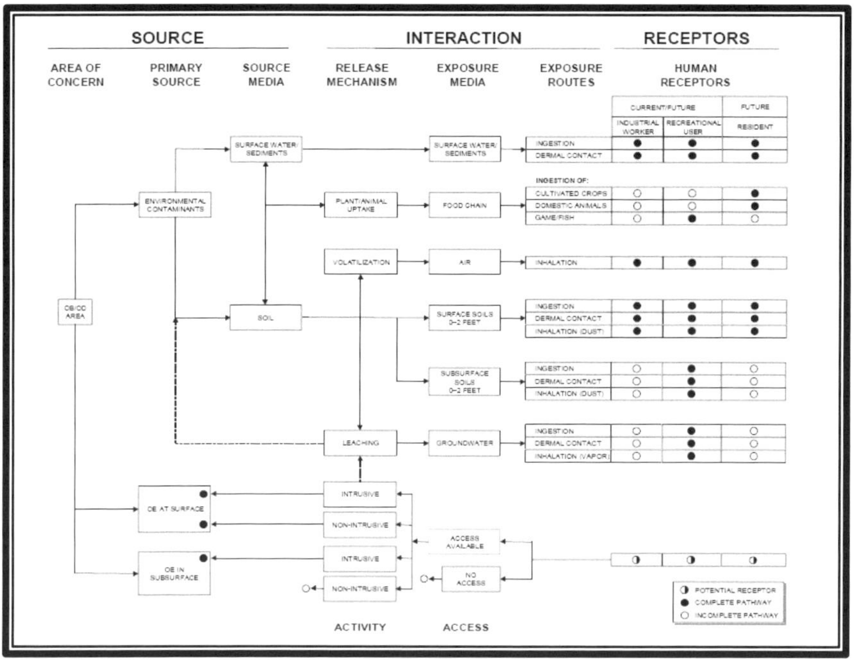

Figure 3. Example Ordnance and Explosive (OE) Munition Response Site (MRS) Conceptual Site Model (CSM) for depicting site conditions. MRS CSMs depict the interaction of the ordnance with the human and ecological environment. MRS CSMs differ from typical CSMs by presenting "hazard access" and "site activity" ([11]).

Typically, MEC/ROW (including UXO and landmines) recovery data is recorded using a variety of methods. "Pen and paper" data collection methods are typically used, with site-specific electronic data entry becoming more popular. Central compilation and evaluation of MEC/ROW recovery datasets across MRS for improved conceptual site modeling is extremely difficult because of these paper-based datasets and non-standardized electronic data sets. However, emerging tools and technologies can be applied to improve MRS CSMs and respond to the MEC/ROW threats.

Two tools exist to potentially help save time and money in collecting, managing, and evaluating munitions recovery data (e.g., for UXO and explosive ROW) at MRS within the Balkan Region: the UXO Recovery Database (RDS) and the Electronic Data Collection Tool (EDCT) ([13,14]). Proper application of these tools within the region can dramatically increase CSM accuracy and associated response.

The UXO RDS was developed to understand the true nature of UXO from the evaluation of data across varied MRS. It offers site managers a centralized, electronic repository for MEC/ROW recover data from multiple MRSs. In addition, it allows for the storage of specific site conditions including topographic features, vegetative features, and soil types. It stores data electronically for multiple MRS with a basic expandable capability. It also allows users to perform data searches and statistical analyses for response decision making. The UXO RDS produces for communication results in tabular, statistical, and graphical formats. It includes components for data entry, quality assurance, and administration thereby providing mechanisms for data control, updates, expansion, and maintenance. The EDCT is a handheld PDA loaded with software developed specifically to handle MEC recovery data allowing data collected in the field to be uploaded directly to the UXO RDS. It includes a Global Positioning System (GPS) for locating MEC/ROW ordnance pieces as well as digital photograph capability for storing a pictorial record of ordnance recovery. The EDCT is operated with drop-down menu navigation and offers synchronization with the UXO RDS from any PC connected to the Internet.

Using tools such as these in the Balkan Region could essentially eliminate the use of *pen and paper* data collection methods as well as help ensure consistency and the use of standard nomenclatures for data collection across geographical boundaries. Most importantly, combining computerized tools with MEC CSM with direct site application would allow for the improvement of conceptual site modeling for MRS response areas. For example, by applying these types of tools, site managers can obtain MEC/ROW penetration or emplacement characters for a given MRS, or for extrapolation to multiple MRS, when considering site-specific conditions. In addition, users could develop the capability to more accurate and rapidly evaluate and summarize MEC/ROW site information which would allow for more effective conceptual site modeling and improved approaches to establish MEC/ROW recovery procures for restoration projects.

System Case Study – Notional MEC/ROW Removal Site

Key to improving response within the Balkan Region is the application of MEC/ROW recovery tools in building improved MEC CSMs for protecting public health. In short, application of the UXO RDS and associated EDCT can enhance the accuracy in estimation of the minimum and maximum intrusive depths for subsurface MEC recovery thereby effectively targeting response depth excavation. To demonstrate application and the development of an improved data points within a MEC CSM, a system case study was developed with a notional removal site.

For the purposes of the exercise, it was assumed a site existed with basic contamination that included mortars and projectiles. The assumed sizes for the MEC were 105mm, 120mm, 155mm, 60mm, and 81mm high explosive (HE) ordnance. Analyzing the data within the UXO RDS [9], a sample size of 418 data points were identified from multiple MRS (CONUS and OCONUS) (Fig. 4). Soil type across the recovery data points was found to be silty-sand clay (68%), unknown (23%), silt loam (5%), and rocky soil type (4%). The minimum depth from the data housed in the RDS was found to be surface contamination. The 95% confidence interval for recovery depth of this family of ordnance pieces was found to be 21.3 inches. Other statistical information for the ordnance pieces for this family of HE included: measures of central tendency such as the mode at 6.0 inches, the mean at 7.1 inches. The standard deviation was found to be 7.8 inches. The 99% confidence interval for recovery was 26.2 inches and the maximum data point was 54 inches. As part of the exercise, an improved notional MEC CSM was developed that includes these data points showing that 95% confidence interval for recovery is approximately 21.3 inches (Fig. 5). Essentially, establishing a response plan that plans for excavation depths to the amounts identified within the notional CSM should eliminate a large majority of the risk from exposure to HE ordnance pieces at a similar site with MEC from this category of ordnance. MRS site managers would also know from this detailed CSM that a small risk does exist for finding ordnance at depths from 21.3 to 26.2 inches with potential for statistical outliers beyond that depth. Excavations beyond approximately 26 inches would have to balances risk versus reward considering the many different site-specific aspects of a site.

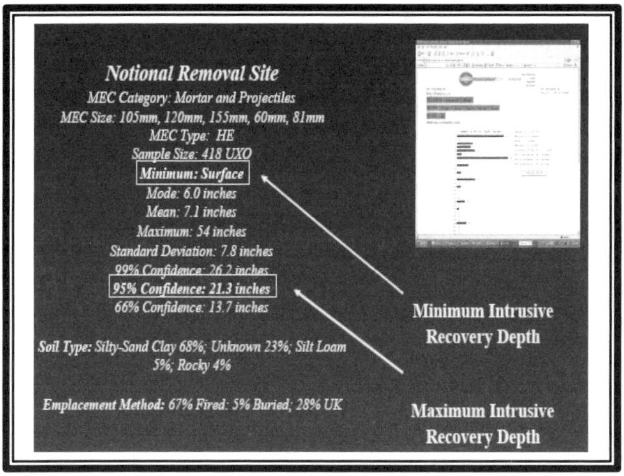

Figure 4. Data set for a Notional MEC/ROW Removal Site identifying minimum and maximum intrusive recovery depths for UXO ([13]).

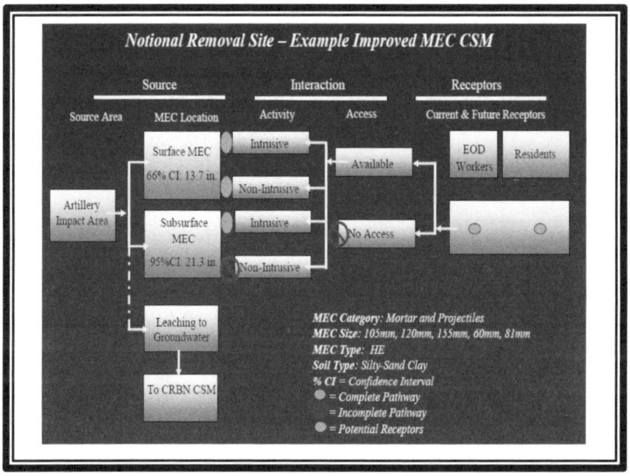

Figure 5. Improved MEC CSM for a notional removal site. Essentially all pieces for the selected high explosive ordnance family could be found within 21.3 inches (95% Confidence Interval) (Fig. 5 data source: [13]). This type of site modeling can save time and money and improve public health response operations at MRS within the Balkan Region.

Summary

Tools now exist to improve the cost effectiveness and timeliness of response for eliminating the risk from MEC/ROW hazards within areas such as the Balkan Region. This exercise demonstrated that CSM can be markedly improved by including identified minimum and maximum UXO intrusive depths, which can thereby further target excavation and recovery efforts for MEC/ROW and identifying the risks at various depths.

It is believed that with the basic, easily applied computer tools, improved MEC CSMs and focused UXO recovery efforts can become a critical element of the process for improving public health preparedness in the Balkan Region where these explosive and eco-hazardous regions exist. The benefits realized from application of computerized tools combined with consistent data collection methods for MEC/ROW recovery areas across the Balkan Region can result in the following:

- Improved CSMs can be developed that present the understanding of the true nature of MEC and ROW (e.g., penetration or emplacement characteristics).
- Improved understanding of the physical setting and release profiles associated with MEC/ROW as well as mechanism of release and exposure profiles to prevent public exposures.
- Improved ability to conduct "what-if" analyses as well as rapidly evaluate and summarize MRS site information.
- Effective approaches to establishing procedures for MEC/ROW restoration projects.

Although there are many benefits, challenges and limitations remain. Areas of continued development and improvement include, but may not be limited to, instituting the use of electronic data collection and management devices, promoting the consistent use of nomenclature, and data quality assurance and application across politically and socially diverse areas. Nonetheless, the low-cost implementation of a data housing tool such as the UXO RDS and associated EDCT with associated benefit dissemination in the Balkan Region can show improved response efficiencies and improved public health preparedness through MEC/ROW risk reduction at high-impact areas.

References

[1] Fiscal Year 2003 Defense Environmental Restoration Program Annual Report to Congress. Office of the Under Secretary of Defense. The Department of Defense. (2004). Washington, D.C. Available Online at: http://www.dtic.mil/envirodod/DERP/DERP.htm.

[2] Balkan Topographic Map. Wikipedia, 2006. The Free Encyclopedia. Available Online at: http://en.wikipedia.org/wiki/Balkans. Visited October 10, 2006.

[3] Portfolio of Mine Action Projects, Ninth Edition. United Nations Mine Action Service, 2006. U.N. Mine Action Service (Department of Peacekeeping Operations), U.N. Development Programme. United Nations Children's Fund. United Nations. New York, NY.

[4] Personal Communication Regarding MEC Contamination in the Balkans, between Taryn Burul, Landmine Monitor Project Support Officer, Mines Action Canada (http://www.icbl.org/lm/), and Mr. Chuck Tomljanovic, University of Pittsburgh, June 2006.

[5] The Bosnia and Herzegovina Mine Action Center, 2006. Bosnia and Herzegovina Humanitarian Demining Operational Plan for 2006. http://www.bhmac.org/.

[6] Arnold, A., 2006. Personal Communication Regarding MEC Contamination in the Balkans, between Mr. Alan Arnold, Geneva International Centre for Humanitarian Demining. Geneva, Switzerland, and Mr. Chuck Tomljanovic, University of Pittsburgh, June 2006.

[7] Standard Guide for Developing Conceptual Site Models for Contaminated Sites. Designation: E 1689 – 95 (Reproved 2003). ASTM International. (2003). West Conshohocken, PA.

[8] Ecological Risk Assessment Guidance for Superfund. U.S. EPA, 2006. Available Online at: http://www.epa.gov/. Example Conceptual Site Models (Fig. 2) from U.S. EPA site: http://www.epa.gov/R5Super/ecology/html/erasteps/erastep3.html#csm. Visited October 10, 2006.

[9] Superfund. Available Online at: http://www.epa.gov/superfund/. Site Visited October 10, 2006. U.S. EPA, 2006b.

[10] MacDonald, J., et al. 2004. Unexploded Ordnance: A Critical Review of Risk Assessment Methods. RAND Corporation. Santa Monica, CA.

[11] Conceptual Site Models for Ordnance and Explosives (OE) and Hazardous, Toxic, and Radioactive Waste (HTRW) Projects. U.S. Army Corps. of Engineers, 2003. Document EM 110-1-1200. Department of the U.S. Army, Washington, DC.

[12] Personal Communication Regarding MEC Contamination in the Balkans, between Mr. Walter "Keith" Angles, U.S. Army Corps. of Engineers (U.S. ACE), Military Munitions Response Team, U.S. ACE UXO Center of Excellence, Huntsville, AL, and Mr. Chuck Tomljanovic, University of Pittsburgh, May 2006.
[13] Limited Access UXO Recovery Data Analysis Database. UXO Recovery Database. (2005). National Defense Center for Environmental Excellence (NDCEE). Available online through the World Wide Web: http://uxords.ctcgsc.org/admin/login.asp. Visited October, 2006.
[14] The UXO Recovery Database and EDCT. National Defense Center for Environmental Excellence (NDCEE), 2006. http://uxords.mvr.usace.army.mil/admin/login.asp.

6. Biological Agents

Strengthening National Public Health Preparedness and Response to Chemical, Biological and Radiological Agent Threats. Edited by C.E. Cummings and E. Stikova.
IOS Press, 2007.

Crimean-Congo Hemorrhagic Fever – A Biological Weapon?

Etem AKBAS*
Mersin University Faculty of Medicine,
Department of Medical Biology and Genetics, Turkey

Abstract. The aim of this report is to develop consensus-based recommendations for measures to be taken by medical and public health professionals if Crimean-Congo hemorrhagic fever (CCHF) viruses are used as biological weapons against a civilian population. CCHF is a viral haemorrhagic fever of the *Nairovirus* genus, family *Bunyaviridae*. CCHF virus is a tick-borne virus that causes a severe hemorrhagic disease in humans with a case fatality rate of approximately 30%. Although primarily a zoonosis, sporadic cases and outbreaks of CCHF affecting humans do occur. The disease is endemic in many countries in Africa, Europe and Asia. Recently, Turkish populations have suffered from Crimean – Congo Hemorrhagic Fever about four years. CCHF has been seen some of the cities in Turkey since 2002. Recommendations for limiting the spread of CCHF, in cases of natural outbreaks or terrorist spread, include.

Keywords. Congo-Crimean haemorrhagic fever, Turkey

Introduction

Congo-Crimean haemorrhagic fever (CCHF) virus is a tick-borne virus that causes a severe hemorrhagic disease in humans with a case fatality rate of approximately 30%. CCHF virus (CCHFV) has been reported in the Near, Middle, and Far East countries, including Iraq, Pakistan, United Arab Emirates, Kuwait, Oman, China and Kosovo [1–5]. CCHFV is a member of the *Nairovirus* genus of the family *Bunyaviridae* [6]. CCHF was first observed in the Crimea by Russian scientists in 1944 and 1945. Congo virus was first isolated in Africa from the blood of a febrile patient in Congo in 1956. For this reason it was named: Crimean – Congo Hemorrhagic Fever. CCHFV is transmitted to humans by *Hyalomma* ticks or by direct contact with the blood of infected humans or domestic animals. The virus is geographically restricted to the areas where their host species of ticks live. CCHF can be transmitted from one infected human to another by contact with infectious blood or body fluids. Documented spread of CCHF has also occurred in hospitals due to improper sterilization of medical equipment, reuse of injection needles, and contamination of medical supplies [7].

* Corresponding Author: Etem AKBAS, Mersin University Faculty of Medicine. Department of Medical Biology and Genetics, Mersin/Turkey; e-mail: etem_a@yahoo.com.

Figure 1. *Hyalomma spp* [8].

1. Biology of *Nairovirus*

Nairovirus include RNA as genetic material, with helical capsid and viral envelopes. Incubation period is 1–3 days after *Hyalomma* bites. An infected human's secretions and/or body fluids contain virus after 5–6 days, and a victim's condition may be worst in about 12 days. Numerous genera of ixodid ticks serve both as vector and reservoir for CCHFV; however, ticks in the genus *Hyalomma* are particularly important to the ecology of this virus. In fact, occurrence of CCHF closely approximates the known world distribution of *Hyalomma* ticks (Fig. 1). Therefore, exposure to these ticks represents a major risk factor for contracting disease [6]. Numerous wild and domestic animals, such as cattle, goats, sheep and hares, serve as amplifying hosts for the virus [7]. *Hyalomma* families include 866 species, of which only 31 species are known to transmit *Nairovirus* [8].

1.1. Symptoms of CCHF

The onset of CCHF is sudden, with initial signs and symptoms including headache, high fever, back pain, joint pain, stomach pain, and vomiting. Red eyes, a flushed face, a red throat, and petechiae (red spots) on the palate are common. Symptoms may also include jaundice, and in severe cases, changes in mood and sensory perception. As the illness progresses, large areas of severe bruising, severe nosebleeds, and uncontrolled bleeding at injection sites can be seen, beginning on about the fourth day of illness and lasting for about two weeks [9,10].

1.2. Prognosis and Therapy

In sheep and cattle the virus has only a slight effect. In humans, the disease can be much more severe. Infected persons can die in two weeks (two week mortality). Treatment options for CCHF are limited, as we have no proven therapy for CCHF. Immunotherapy and ribavirin have been tried with varying degrees of success during sporadic outbreaks of disease, but no case-controlled or cohort trials have been conducted [9,10].

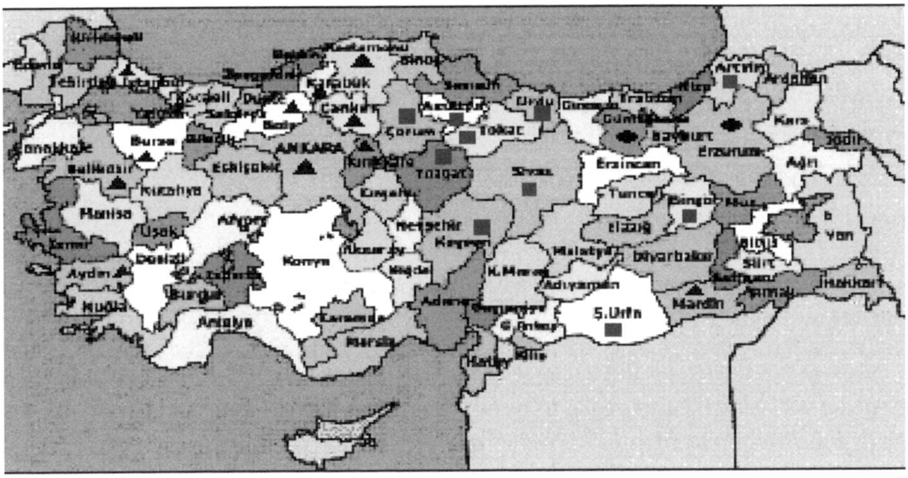

: 2002-2003 : 2004-2005 : 2006

Figure 2. CCHF in the cities of Turkey, 2002–2006.

1.3. Protection and Control Measures

Persons in high-risk occupations slaughterhouse workers, veterinarians, sheep herders, etc., should take every precaution to avoid exposure to virus-infected ticks or virus-contaminated animal blood or other tissues. For example, wearing gloves and limiting exposure of naked skin to fresh blood and other tissues of animals are effective practical control measures. Likewise, medical personnel who care for suspected CCHF patients should practice standard barrier-nursing techniques. Tick control may not always be practical in many regions of the world where *Hyalomma* ticks are most prevalent. However, acaricide treatment of livestock in CCHF endemic areas is effective in reducing the population of infected ticks [9,10].

2. Crimean-Congo Hemorrhagic Fever in Turkey

In Turkey, CCHF cases were first seen in Gümüşhane and Erzurum in 2002 and 2003 [9,11]. In 2004 and 2005, it began to spread both in the direction of eastern cities Artvin and Bingöl and western cities Sivas, Erzincan, Tokat, Amasya, Çorum, Ordu, Yozgat, Kayseri and Şanlı Urfa. In 2006, CCHF cases became more intense in Çorum, Yozgat, Amasya and Tokat, and were now seen in Ankara, Artvin, Aydın, Balıkesir, Bolu, Çorum, Erzurum, Gümüşhane, Karabük, Kastamonu, Kırıkkale, Mardin, Sivas, Bursa and İstanbul (Fig. 2) [11].

According to the Ministry of Health figures, 815 cases were reported from 2002 to 29 June 2006 and 43 cases resulted in death (Table 1). With respect to the past four years, during which the cumulative number of cases (and deaths) was 150 (6) in 2003; 249 (13) in 2004, 266 (13) in 2005; and 150 (11) in 2006 is suggestive of increased disease activity [11].

Table 1. The distribution of CCHF cases and mortal cases according to years in Turkey

Years	Cases	Deaths
2002–2003	150	6
2004	249	13
2005	266	13
2006 (June 29)	150	11
Total	815	46

In Turkey, as diagnosed CCHF cases and the cities with mortal cases by CCHF are studied, an orderly geographical distribution is seen. First, it was seen in the cities of Kelkit valley district of Erzurum, Gümüşhane, Sivas and Tokat. In the following years, the neighboring eastern and western cities were regularly added to the above- mentioned cities. CCHF cases began to be seen starting in May, increased in June-July and started to decrease in August for recent last four years.

3. Recommendations

The Ministry of Health of Turkey, in close collaboration with the Ministry of Environment and the Ministry of Agriculture and Rural Affairs, is currently implementing control measures, and enhanced CCHF surveillance has been established nationwide. Case management guidelines, including a treatment protocol with ribavirin, have been distributed to health care facilities throughout Turkey. Four referral hospitals with isolation facilities have also been designated in Ankara, Erzurum, Sivas and Samsun. Public awareness campaigns are ongoing, stressing the adoption of personal protective measures to avoid tick bites, and targeting the rural population through television, radio, posters and leaflets. People in the occupational categories most at risk have been alerted and informed about personal protective measures. These same measures can be applied to outbreaks in other countries and to intentionally spread outbreaks, such as terrorist events.

References

[1] A.S. Altaf, A.J. Luby, N. Ahmed, A.J. Zaidi, S. Khan, J. Mirza, McCormick, and S. Fisher-Hoch. Outbreak of Crimean-Congo haemorrhagic fever in Quetta, Pakistan: contact tracing and risk assessment. Trop. Med. Int. Health 3 (1998), 878–882.

[2] L.L. Rodriguez, G.O. Maupin, T.G. Ksiazek, P.E. Rollin, A.S. Khan, T.F. Schwarz, R.S. Lofts, J.F. Smith, A.M. Noor, C.J. Peters, S.T. Nichol. Molecular investigation of a multisource outbreak of Crimean-Congo hemorrhagic fever in the United Arab Emirates. Am. J. Trop. Med. Hyg. 57 (1997), 512–518.

[3] E.M. Scrimgeour, A. Zaki, F.R. Mehta, A.K. Abraham, S. Al-Busaidy, H. El-Khatim, S.F. Al-Rawas, A.M. Kamal, A.J. Mohammed. Crimean-Congo haemorrhagic fever in Oman. Trans. R. Soc. Trop. Med. Hyg. 90 (1996), 290–291.

[4] Y.C. Yen, L.X. Kong, L. Lee, Y.Q. Zhang, F. Li, B.J. Cai, S.Y. Gao. Characteristics of Crimean-Congo hemorrhagic fever virus (Xinjiang strain) in China. Am. J. Trop. Med. Hyg. 34 (1985), 1179–1182.

[5] C. Drosten, D. Minnak, P. Emmerich, H. Schmitz, T. Reinicke. Crimean-Congo hemorrhagic fever Kosovo. J. Clin. Microbiol. 40 (2002) 1122–1123.

[6] S.T. Nichol, Bunyaviruses, 2001 p. 1603–1633. *In* D.M. Knipe and P.M. Howley (ed.), Fields virology, 2nd ed. Lippincott Williams & Wilkins, Philadelphia, Pa.

[7] Ö. Ergönül, A. Celikbas, B. Dokuzoguz, *et al.* "The chacteristics of Crimean-Congo hemorrhagic fever in a recent outbreak in Turkey and the impact of oral ribavirin therapy". Clin Infect Dis 39 (2004), 285–89.
[8] E. Yalçın. Hayvanlardan insanlara geçen hastalıklar: Kırım Kongo Kanamalı Ateşi. Erzurum Veteriner Kontrol ve Araştırma Enstitüsü Müdürlüğü Yayını. Erzurum, 2003.
[9] C.J. Peters. Viral Hemorrhagic Fevers. *Viral Pathogenesis*. New York: Lippincott-Raven Publishers, (1997), 779–794.
[10] C.A. Whitehouse. Crimean–Congo hemorrhagic fever. Antiviral Research 64 (2004) 145–160.
[11] http://www.kirim-kongo.saglik.gov.tr/.

7. Chemical Agents

Strengthening National Public Health Preparedness and Response to Chemical, Biological and Radiological Agent Threats. Edited by C.E. Cummings and E. Stikova.
IOS Press, 2007.

Fundamentals of Preparedness against Chemical Threats – Introduction

Alessandra ROSSODIVITA[a], Elisaveta Jasna STIKOVA[b], Curtis E CUMMINGS[c]
[a]*Department of Cardiothoracic and Vascular Diseases, San Raffaele Hospital Scientific Foundation, University "Life and Health", Milan, Italy*
[b]*University St Cyril and Methodius Faculty of Medicine, Department of Environmental & Occupational Health, Skopje, Rep of Macedonia*
[c]*Drexel University School of Public Health, Philadelphia, PA, USA*

Abstract. To introduce the topic of chemical agents, we discuss their place in the context of terrorist use of weapons of mass destruction, their recent uses on human populations, and the approach to national and international response. Despite international treaties and efforts to control the spread of chemical agents, these weapons remain a real threat. Both chemical agents and industrial hazardous materials can be used by terrorists. The public health system must be ready for chemical attack by having proper plans, infrastructure and supplies, and public health surveillance systems all in place. Improved and increased research in medical countermeasures, information sharing and coordination, education and training programs represent the key words to manage this emerging priority.

Keywords: chemical warfare, public health preparedness, hazardous materials and terrorism, public health surveillance

Background

This chapter introduces the ASI section on chemical agents, presents guidelines for these agents and suggests what is fundamental to create civil and military defence systems against the most likely warfare and terror threats. The relative ease of production and dissemination of chemical warfare agents – as well as the fact that they have actually been used against populations in the past 20 years – makes chemical terrorism the most likely type of terrorism in the future after bombings and multiple shootings. The possibility of terrorist attack involving the use of toxic chemical substances – industrial or weapons of mass destruction (WMD) – is a world-wide concern for public authorities.

Scope

The ultimate terrorist threat is often seen as the possible use of WMD – chemical, biological, radioactive/nuclear and explosives (CBRNE), although the use of these weapons has not been frequent in the past. The public health threat of terrorism using weapons of mass destruction (WMD-T) has become increasingly real.

As is well-known, terrorism is designed to generate fear, to intimidate, to affect government conduct, to punish a specific target, to kill or maim as many people as possible, to gain news media attention, and to undermine the social stability. Terrorist attacks are unique among disasters in that they are intentional. The terrorist organizations have become better organized, better financed and have increased access to WMD. Developments in transportation, telecommunications and information, and easy access to scientific studies have greatly facilitated the work of terrorists and their organizations.

Although a chemical agent attack may kill fewer people than a nuclear weapon or a massive anthrax spore release, the word "sarin" by itself provokes great fear, and the likelihood of chemical use is significant.

Resurgence and recent use

During the last decade terrorism-related emergencies and conflicts have had a resurgence causing a complex pattern of world-wide changes and imbalance. Throughout the globe, healthcare providers are increasingly challenged by the spectre of terrorism and the threat of WMD. The terrorist events of the past few years, highlighted by the attacks on the World Trade Center and the Pentagon (11/09/2001), Madrid and London placed added risk on the table, changing forever the mindset of antiterrorism responses. Terrorist acts have ranged from dissemination of aerosolized anthrax spores, to intentional food product contamination, release of chemical weapons in major metropolitan subway systems, and suicide attacks using explosives [1,2,3,4,5,6,7].

Chemical warfare agents differ from other WMD-T agents because of the extent of their past use. They remain a continuing problem despite international treaties and efforts to control proliferation of weapons, because of their availability. Despite the Geneva protocol (1925) and the Chemical Weapons Convention (1993) that banned the development, production, stockpiling and use of chemical warfare agents, in recent years several countries have produced, attempted to produce or used chemical weapons, not only in conflicts with other countries but also in domestic conflicts against their own populations [8,9,10]. One example was the Iraqi military force that deployed nerve agents and mustard against both the Iranian army and against several Kurdish villages (1984-1988). Then, in 1995 the sarin attack on Tokyo subway trains caused more than 5.500 casualties with 13 people killed, demonstrating that a well-financed terrorist organization can produce and deploy a chemical WMD using common chemical precursors and manufacturing technology [11, 12].

Risk of both chemical agents and hazardous materials

Chemical warfare agents are generally thought to be compounds with unique toxicity, designed and produced for military use. However, this is only partly true. Nerve agents are chemical relatives of commonly-used organophosphate insecticides. The agents phosgene, chlorine and cyanide were used in chemical warfare during World War I but also are widely used industrial chemicals. Botulinum toxin and mustard have indications as therapeutic drugs. It is not easy to establish clear limits between chemical warfare agents and other highly toxic compounds [13, 14].

In past years, chemical WMD's, e.g., nerve agents or blister agents, were considered the main threat of terrorist chemical assault, but today it is recognized that terrorist use of industrial chemicals as a weapon poses another non-conventional risk. Civilian population are at risk from accidental or deliberate exposure to Hazardous Materials (HazMat) used in everyday life, and the potential threat may be equal to that of chemical agents if used for a terrorist act. The spectrum of HazMat that might be involved in a toxicological incident has increased. Both chemical WMD's and industrial HazMats are characterized by qualities of toxicity, latency of action, persistency and transmissibility [15,16,17,18,19,20].

Response to chemical threats

Saving lives and minimizing the health risks are the first and most important challenges for emergency medical services (EMS). Therefore, prevention of mortality is critically dependent on hospital preparedness at the local and national level. Medical care of the victims will depend on timely intervention, availability of equipment and trained experts as well as effective field detection devices and specific laboratory analytical procedures. Unfortunately, hospital preparedness for chemical disasters is often reported to be inadequate.

The world has developed a heightened awareness of the security threat posed by terrorists and now understands that terrorists are continually innovative in planning new terrorist operations. The prediction of a possible future terrorist attack drives the efforts of public policy and government funding. It is well-known and crucial to underline, however, that there is no generic protection against all possible chemical warfare agents. Many threat-analysis scenarios have been performed, and there is no clear answer as to which chemical agents are the most likely threats in chemical warfare or from chemical terrorism. Therefore, a comprehensive approach is required.

Recommendations

NATO members and Eligible Partner nations must take steps to improve their preparedness for chemical threats, both individually and by creating a NATO-wide system. ASI participants concur with the following steps:

Plans

Preparedness for attacks using chemical agents has become an important issue for national and international security programs in terms of risk reduction strategies, government plans, installation plans, medical and health institution preparedness and research programs. Each nation's EMS should adapt HazMat plans and training to deal with the new threat; there are many similarities between the treatment of casualties from chemical warfare and HazMat agents. In addition, it is essential to maintain military and civilian healthcare providers who have the knowledge and the ability to early and rapidly recognise chemical warfare exposure symptoms, as well as the necessary medical and drug support. Hospital emergency plans that describe specific procedures, precautions and personal protective equipment (PPE) used during chemical exposures or during HazMat releases are required. Therefore, enhanced and coordinated international medical research efforts represent the cornerstone necessary to improve the effectiveness of both medical prophylaxis as well as the therapy.

The global fight against terrorism requires efforts in many areas and directions, from strengthened border controls to better coordination among numerous government agencies at the local, national and international levels. The use of expert panels of scientists and the expertise of local, state and federal and international partners can provide additional security.

Infrastructure and supply

The medical and public health infrastructure and supply systems must be prepared to prevent and treat illness and injury that could result from CBRNE terrorism. In the health sector, preparedness for such situations is much the same as for other catastrophic events. A common civil and military defense program should include the following:

- Proper information and analysis systems;
- Stockpiles of PPE, antidotes and decontamination equipment;
- Detection equipment and analytical means to determine the presence of chemical warfare agents;
- Research programs on medical countermeasures to national and international levels.

Public health surveillance

Preparedness for terrorist-caused outbreaks and injuries is an essential component for every nation and for every public health system, which is designed to protect the population against any unusual public health event. Preparing the nation to address these dangers is a major challenge to public health systems and health-care providers. The fundamental issues are: preparedness and prevention, detection and surveillance, diagnosis and characterization of chemical and biological agents, response and communication. To be effective, an integrated training program must also include disaster drills, and address management and response plans to ensure core competency in public health preparedness.

Conclusions

The international community today is faced with multifaceted and multidirectional risks, including both an increased potential of regional conflicts as well as the proliferation of WMD and terrorism. These challenges drive new priorities in terms of solidarity, and common responses that are based on jointly defined norms.

Preparedness plans must include both the military and civilian public health and healthcare systems, and address both chemical WMD agents and industrial HazMat. Proper infrastructure and supply systems are required, including ample stockpiles of PPE and antidotes. The authors would underline the fundamental issue of common cooperation between civil and military defense program in terms of security, medical preparedness, prevention, research programs and public health surveillance. To improve and increase research in medical countermeasures, information sharing, coordination, education, training programs and security with the cooperation of a highest panel of scientists and expertise at the local, national and international levels – these are the key words to manage this emerging priority.

References

[1] EK Noji. Scientific Networking and the Global Health Network Supercourse (2005) 3
[2] EK Noji. *Prehospital Disaster Medicine*; 18 (3) (2003) 163
[3] EK Noji. The public health consequences of disasters. New York: Oxford University Press (1997)
[4] EK Noji. *International Journal of Disaster Medicine*, 1 (2003) 1
[5] D Alexander. *Prehospital Disaster Medicine*; 18 (3) (2003) 165
[6] C DiGiovanni. *Prehospital Disaster Medicine;*18 (3) (2003) 253
[7] ME Keim, N Pesick and N.A.Y. Twum-Danso. *Prehospital Disaster Medicine;* 18 (3) (2003) 193
[8] W Turnbull, P Abhayaratne. 2002. Monterey Institute of International studies: Center for Non-proliferation Studies, Washington, D.C. (2003)
[9] M Schwenk, S Kluge, H Jaroni. *Toxicology* 214 (2005) 232
[10] S. Gahlaut and G. K. Bertsch, eds. *Orbis*, Elsevier, 48 (3) (2004) 489
[11] P Aas. *Prehospital Disaster Medicine;* 18 (3) (2003) 208
[12] JL Arnold. *Prehospital Disaster Medicine*; 17 (1) (2002) 3
[13] T Zilker. *Toxicology* 214 (2005) 221
[14] L Szinicz. *Toxicolgy* 214 (2005) 167
[15] A.Krivoy et al. *Prehospital Disaster Medicine;* 20 (3) (2005) 155
[16] J Borak, M Callan, W Abbott. New Jersey: Prentice Hall (1991)
[17] D Baker. *Prehospital Disaster Medicine* 19 (2) (2004) 174
[18] World Health Organization, *Public health response to biological and chemical weapons: WHO guidance.* Geneva, 2004
[19] AC Bronstein, PL Currance. Emergency Care for Hazardous Materials Exposure, Mosby Lifeline 2nd Ed., 1994
[20] DJ Baker. *Current Anaesth Crit Care* 9 (1998) 52

Strengthening National Public Health Preparedness and Response to Chemical, Biological and Radiological Agent Threats. Edited by C.E. Cummings and E. Stikova.
IOS Press, 2007.

Chemical Warfare Agents – Medical Aspects and Principles of Treatment, Part I – Nerve Agents

Alessandra ROSSODIVITA [a,1], Matteo GUIDOTTI [b], Massimo C. RANGHIERI[c], Elisaveta Jasna STIKOVA [d], Curtis E. CUMMINGS [e]

[a]*Department of Cardiothoracic and Vascular Diseases, San Raffaele Hospital Scientific Foundation; University "Life and Health", Milan, Italy,*
[b]*CNR-Institute of Molecular Sciences and Technologies, Milan, Italy*
[c]*SMOM Auxiliary Corps of the Italian Army, 1° Reparto, Milan, Italy*
[d]*University "St Cyril and Methodius" Faculty of Medicine; Skopje, Republic of Macedonia*
[e]*Drexel University School of Public Health, Philadelphia, PA, USA*

Abstract: Despite international treaties with strong verification measures that aim to prohibit and prevent the use of weapons of mass destruction, nevertheless some countries and terrorist groups have developed, produced and used such weapons. Successful management of casualties from chemical warfare agents strongly depends on fast and appropriate medical treatment, on the ability of first responders to take proper action, and on new and more effective counter-measures. Although the general principles of clinical toxicology, such as decontamination, stabilization, patient evaluation and symptomatic treatment are similar for many toxicants, chemical warfare agents deserve special attention because their very high toxicity, rapid onset and multiple organ involvement with lethal evolution. This report describes briefly the toxicity and the medical management of mass casualties with chemical warfare agents. Characteristic diagnostic signs and therapeutic schemes for these agents are described. The importance of the knowledge of medical and emergency personnel as well as that of military nuclear, biological or chemical (NBC) unit preparedness and improving research in the field of chemical warfare are emphasized.

Keywords: Chemical warfare agents, terrorism, toxicology, medical treatment, preparedness

1. Chemical and toxic agents – definition and risks

The use of chemical warfare agents against civilians and unprotected troops in international conflicts or by terrorists is considered to be a real threat, particularly after the terrorist attacks on the Tokyo subway in 1995 and the World Trade Center and Pentagon attacks on 11 September 2001. This report discusses the mechanism of action of selected chemical warfare agents and the management principles of incidents

[1] Corresponding author: San Raffaele Hospital Scientific Foundation, University " Life and Health," Department of Cardiothoracic and Vascular Diseases, 60 Olgettina Street, 20132 Milan, Italy; E-mail : rossodivita.alessandra@hsr.it

with chemical agents. The authors provide an overview of the clinical aspects of the major chemical weapons with special attention on military chemical agents.

A "chemical threat" can be defined as "a statement of the who, what, where, and how of chemical warfare" [1]. The threat may involve single or multiple chemical agents, not only the classical chemical agents, but also highly toxic substances or industrial compounds that could achieve the same objective. Chemical warfare agents need not be lethal to be disruptive. Chemical and biological warfare agents usually are classified as weapon of mass destruction, but really they are agents of mass injury. Offensive use of chemical agents continues to be attractive for some nations and terrorists, since these agents can be dispersed over large areas and can eventually penetrate even the most well-defended positions.

Civilian populations are at risk from accidental or deliberate exposure to a wide range of toxic substances, both industrial and military [1]. Hazardous Materials (HazMat) represent an emerging threat, also. Civilian emergency medical responders should be able to respond to a possible toxic attack, by adapting their plans for dealing with casualties from HazMat to address a possible chemical-biological, radioactive/nuclear and explosive (CBRNE) threat.

A toxic agent may be defined as any substance that is injurious to health in an unconfined state [2]. The civilian HazMat system classifies many thousands of toxic hazards that are in general production and use [3, 4, 5, 6]. Their toxicity to humans is incidental to their primary purpose, unlike military hazards that have been developed specifically as weapons, and therefore, specifically to cause harm. Military toxic agents have been recognized for more than one hundred years and conventionally have been divided in chemical and biological threats [7].

Chemical agents and HazMat substances may share common pathophysiological effects as well as subsequent organ failure. All toxic hazards in the chemical warfare spectrum may be described according to four major properties:

1. Toxicity
2. Latency of action (1 and 2 are determined by toxicokinetics and dynamics)
3. Persistence
4. Transmissibility (which are determined by the physicochemical properties).

Transmissibility is sometimes referred to as secondary contamination. Toxicity and latency define the danger to the victim and persistence and transmissibility define the danger to emergency medical system (EMS) rescuers [8].

Military chemical agents are characterized according to several features. Among them are the nature of their use, their persistence in the field, and their physiological action. They can produce incapacitation, serious injury and death. Release of a military toxic agent usually takes place as a part of battle; the agents often are identified beforehand by intelligence reports and may be detected by a detection system. The response to attack and injury usually is a part of a structured response to the hazards and threats of the battlefield [8].

Civilian toxic releases are usually accidental and unforeseen. An exception is use in a terrorist attack. Exposure to toxic agents may be sudden or insidious. In the initial phase of a chemical mass-casualty, the identity of the toxicant is usually unknown. The hazards usually are listed in HazMat systems and can be identified from code numbers and databases of published information about the hazards and the measures needed for decontamination. There usually are no fixed detectors, and signs and symptoms among victims are usually the first indicators of attack. Exposures may present variable times of latency. Accurate observation of symptoms is essential to get early information about the possible toxic substances involved and to start immediate medical treatment. Effectiveness of medical response differs from military response, since the degree of preparedness of civilian EMS personnel and hospitals varies.

Many of the civilian hazards listed as HazMat are non-persistent and non-transmissible. As such, they pose no contamination or transmissibility risk. This means that rescue teams can begin work upwind of the release without requiring protection or decontamination. Smokes, which fall into this category, are the toxic hazards most likely encountered in civilian life. Other civilian hazards and most military hazards are both persistent and transmissible and pose a major threat to EMS teams. In such cases, full protected entry to the contaminated zone is required.

Management of chemical incidents requires specific knowledge of the toxicants and their biological effects as well as clinical experience in diagnosis and treatment of intoxications. EMS personnel must know both principles of treatment of common medical emergencies, as well as decontamination and personal protection procedures.

Table 1. The major chemical warfare agents [1] according to the U.S. naming code

U.S Army code	Agent
Cyanides	
AC	Hydrogen cyanide
CK	Cyanogen chloride
Nerve agents	
GA (Tabun)	Ethyl N,N-dimethyl-phosphoramidocynidate
GB (Sarin)	Isopropyl-methylphoshonofluridate
GD (Soman)	1,2,2-trimethylpropyl methylphosphonofluoridate
GF	
VX	
Lung toxicants	
CG (Phosgene)	Carbnonyl chloride
DP (Diphosgene)	Trichloromethyl chloroformate
Vesicants	
HD (Mustard)	*Bis*-2-Chloroethyl sulphide
L (Lewisite)	2-Chlorovinyl dichloroarsine
HL	Mustard-Lewisite misture
Incapacitating agents	
BZ	3-Quinuclidinyl benzilate (QNB)
Tear gas	
CN	2-Chloro-1-phenylethanone
CS	2-Chlorobenzalmalononitrile
Vomiting agent	
DM (Adamsite)	10-Chloro-5,10-dihydrophenarsazine

2. Nerve Agents

Nerve agents were first developed in the 1930's primarily for military use and are extremely toxic. Related substances are used in medicine, in pharmacology, and for other purposes, such as insecticides, but they lack the potency of the military agents [1].

Nerve agents are the most toxic of chemical warfare agents. They acquired their name because their target organ is the nervous system. Their basic chemical structure which they all have in common is described by the Schrader formula [10]. The nerve agents belong chemically to the group known as organophosphate (OP) compounds.

2.1. Mechanism of intoxication

Nerve agents act as cholinesterase inhibitors. The common mechanism of action of OP-nerve agents and OP-pesticides is the inhibition of the cholinesterase enzymes by covalent binding to the amino acid serine in the active center of the enzyme [11]. OP toxins block both butyryl-cholinesterase and acetylcholinesterase (AChE).

AChE rapidly hydrolyses the neurotransmitter acetylcholine in the CNS (central nervous system) and the PNS (peripheral nervous system) [11]. Inhibition of AChE leads to an accumulation of acetylcholine resulting in over-stimulation of the whole cholinergic system. The inhibitory potency of OP's towards AChE determines their relative toxicity. Inhibition is initially a reversible process, but OP-inhibited AChE may undergo a secondary reaction – spontaneous dealkylation through alkyl-oxygen bond scission (aging) – and become irreversibly inactivated. Each nerve agent has different aging kinetics and a different half-time of aging, dependent on the structure of each compound (e.g., soman-inhibited AChE ages within minutes while the aging half-time of sarin-inhibited enzyme is 3h). Both speed of onset of signs of poisoning and velocity of aging greatly affect treatment procedure and therapeutic efficacy of antidotes. [12].

2.2. General characteristics and exposure routes

The classical nerve agents are tabun (GA); sarin (GB), cyclosarin (GF), soman (GD) and VX. They may be incorrectly called nerve "gases" – instead, most are liquids at room temperature that readily become vapors. They are stable chemically, are dispersed easily, and are absorbed via inhalation, through the skin and by the gastrointestinal tract. They vary somewhat in their physical and chemical properties and their toxicity. In 1992, some authors described the Soviet-developed "Novichoks," a supposed new class of binary nerve agent, and being binary (two less-active components will combine to form the nerve agent), cannot be detected by existing electronic detectors [13, 14, 15, 16]. In spite of these differences, their toxic action is the same, as described above. Nerve agent vapor is denser than air, thus it remains close to ground. Nerve agents tend to disperse and are classified as "non-persistent" in the environment except for VX.

The main routes of intoxication are inhalation of vapor and direct absorption through the skin. Effects produced by nerve agent vapor begin in seconds to minutes after the onset of exposure; depending on the concentration of vapor, these effects usually reach maximal severity within minutes after the patient is removed from or protected from the vapor or may continue to worsen if the exposure continues.

Nerve agents are easily produced and readily available. Their synthesis is relatively simple and inexpensive and they cause morbidity and mortality at extremely low doses. Some persist in the environment for long periods of time and can be released from contaminated clothing, skin and secretions [17]. G-type agents are clear, colourless and tasteless liquids that are miscible in water and most organic solvents.

Nerve agents like soman, sarin and tabun can affect the eyes, the nose, airways, or a combination of these organs at low concentrations.

2.3. Tabun (GA)

Tabun is characterized by a faint fruity odour, the appearance is colourless to brownish liquid; the half life is 1.5 days; the median lethal dose is:

- LCt_{50}: 150 to 400mg/min/m^3 (in air);
- LD_{50} (skin): 1.000mg/70-kg man ;

2.4 Sarin (GB)

Sarin is the most volatile and is odourless; the appearance is colourless liquid; the half life is 1.5 days; the median lethal dose is:

- LCt_{50}: 70 to 100 mg/min/m^3 by inhalation;
- LD_{50} (skin): 1.700mg/70-kg man ;

2.5 Cyclosarin (GF)

The median lethal dose is:

- LCt_{50}: 35 mg/min/m^3 by inhalation;
- LD_{50} (skin): 350mg/70-kg man ;

Table 2. Nerve agent poisoning – effect of exposure route [1,9,11,17]

Time of exposure	Route	Time to death
1-2 min	**Skin**	1-2 hours
1 min	**Eye or mucous**	10 min
1-5 min Soman	**Food**	0.5-2hr

Table 3. Nerve agent toxicities [1,9,11,17]

Agent	Relative toxicity	Inhalation mean lethal dose (*) (mg/min/m3)	Skin exposure mean lethal dose (mg)
Tabun (GA)	Lowest	150 to 400	1000
Sarin (GB)		75 to 100	1700
Soman (GF)		35 to 50	100
VX	Highest	10	6

Legend: LCt_{50}: median lethal exposure (mg min m^{-3}) in air [9], LD_{50}: lethal dose (dermal) (mg/person)

2.6. Soman (GD)
Soman has a fruity or oil of camphor odour; the appearance is a colourless liquid; it is described as relatively persistent; the median lethal dose is:

- LCt_{50}: 35 to 50 mg/min/m^3 ;
- LD_{50} (skin): 350mg/70-kg man ;

2.7. VX
VX is the most toxic nerve agent, 100 times more toxic than sarin; it is more stable, less volatile and penetrates skin better than other nerve agents; VX is odourless; amber coloured; the usual persistence in the soil is 2 to 6 days, but it can persists for weeks in low temperatures; the median lethal dose is:

- LCt_{50}: 10 mg/min/m^3 ;
- LD_{50} (skin): 10mg/70-kg man ;

All listed nerve agents are detoxified by diluted hypochlorite, soapy water or by use of military decontamination kits [1,9,17,18,19,20].

2.8. Symptoms and effects of nerve agents.

The principal effects of nerve agents are on the eyes, nose, pulmonary system, skeletal musculature, CNS and behaviour, and cardiovascular system. There is also a marked increase in secretions from glands, notably the ocular, salivary, pulmonary and gastrointestinal glands [1,9,11,19].

2.8.1. The eyes. Ocular symptoms occur after local or systemic exposure to a nerve agent. Local nerve agent exposure to eyes causes miosis, conjunctival injection, lachrymation, pain in or around the eye, and blurred or dim vision or both. Reflex nausea and vomiting may accompany eye exposure. These effects can be from direct exposure of the eye to vapor or liquid, while significant systemic exposure to a nerve agent may produce moderate symptoms like nausea and vomiting with or without miosis. The incidence of symptoms varies [9].

2.8.2. The nose. After local or systemic exposure, rhinorrhea is common and can be severe [9].

2.8.3. Pulmonary System. The pulmonary effects can begin within seconds after inhalation. After low-dose exposure to nerve agent vapor, casualties can complain of a tight chest, with difficulty in breathing, due to bronchial muscle spasm. Dyspnea, bronchospasm, bronchoconstriction and apnea represent the four major pulmonary symptoms. Respiratory failure is the primary cause of death [9].

2.8.4. Skeletal musculature. Neuromuscular effects agents are initially expressed as stimulation of small muscle fibers with fasciculations, then by stimulation of muscles and muscle groups with twitches or jerks, finally by fatigue and, at the end-stage, flaccid paralysis. Involvement of respiratory muscles can cause respiratory arrest [9].

2.8.5. CNS and behavior. Neurological symptoms include seizures, slurred speech, mental status changes, ataxia, cerebral ataxia, peripheral neuropathy, lethargy and memory impairment. Exposed casualties can show behavioral and psychological changes. Nervousness, insomnia, tenseness, fatigue, poor comprehension, decrease ability to communicate, mild confusion and depression are reported.

2.8.6. Cardiovascular system. Although cholinesterase inhibitors are classically described to cause severe bradycardia, clinical data show this finding to be inconsistent. Blood pressure may be elevated after a mild-to-moderate exposure. Decreased heart rate with atrial-ventricular (AV) block (I-II-III degree block); transient AV-dissociation and both bradycardia and tachycardia are reported.

3. General treatment principles

Principles of treatment for nerve agent exposure include the same principles used for any toxic substance exposure, as well as few specific available antidotes.

The first priority is to protect the first responders. Oral pyridostigmine and personal protective equipment (PPE) are strongly suggested. Pyridostigmine bromide is a reversible cholinesterase inhibitor, useful as an effective prophylactic measure against nerve agents, particularly against soman; the dosage is 30 mg PO every 8 h.

The keys to treatment of victims are: terminate exposure, remove the casualty from the dangerous area, establish and maintain ventilation (avoid mouth-to-mouth resuscitation), administer antidotes if available, and correct cardiovascular problems.

Initial treatment for exposed casualties is to terminate exposure. This involves clothing removal and decontamination by trained, properly-equipped first responders. Decontamination is achieved by washing with soap and water, or with 0.5% sodium hypochlorite, as stated above; if liquid agent is found on skin or clothing, the agent can be scraped off skin manually before washing (by rescuers or others wearing PPE).

Ventilatory support, if available, is extremely useful and may be necessary, because spontaneous respiration will stop within few minutes after the onset of effects of high exposure to a nerve agent. Ventilation may often be required for 20-30 minutes, up to a maximum of 3 hours.

Atropine is an effective drug to attenuate or eliminate nerve agent effects. Atropine sulphate acts by blocking the effects of excess concentrations of acetylcholine at muscarinic cholinergic synapses. It can reduce morbidity by reversing some or all muscarinic effects, although it is ineffective in controlling the nicotinic effects of nerve agent exposure. The antidote may be self-administered by the victim (by autoinjector - e.g., Atropen) or administered by first responders. Dose is 2 mg of atropine IM or IV at 5-10 min intervals until dyspnea and secretions are minimized. For conscious patients, usually no more than 4 mg will be required, but severely exposed, unconscious casualties have required up to 15 mg to regain consciousness and spontaneous respirations [1].

Oximes – pralidoxime (2-PAM chloride) or obidoxime – reactivate nerve agent-inhibited AChE. Mono- and bis-pyridium oximes constitute an essential and crucial element of post-exposure medical treatment. They are also used in cases of organophosphate pesticide intoxication. To be effective, oximes must be administered within minutes to a few hours after intoxication, depending on the agent involved. The therapeutic dosage of 2-PAM chloride has not been established, but current evidence indicates it to be 15 to 25 mg/kg. The effective dose depends on the nerve agent, the time between exposure and oxime administration, and other factors. Initially, an oxime should be administered with atropine. In cases of severe exposure, 1 to 1.5 g of oxime should be administered IV over a period of 20 to 30 minutes or more [9].

In general, post-exposure therapy following nerve agent exposure includes the use of one or more autoinjectors containing atropine (usually 2 mg) and one of the oximes – obidoxime, P2S, 2-PAM, HI6, or TMB4 [13]. To stop seizures, benzodiazepines are commonly used, especially diazepam, and there is evidence that the drug reduces brain damage [9]. Anticholinergic drugs have proven to be more potent than the benzodiazepines in ending seizures, but anticholinergics must be administered very shortly after the onset of seizures, preferably within the first five minutes following intoxication [9,18,19,20].

4. Future countermeasures

Present medical counter-measures against nerve agents are not sufficiently effective, particularly in protecting the brain [13]. Therefore, new and more effective counter-measures must be developed to enable better medical treatment of civilians and military personnel following exposure to a nerve agent.

New alternative medical countermeasures to nerve agent intoxication are being evaluated continuously, given the emerging threat of these agents by terrorists. Anticholinergics agents, glutamate antagonists (such as NMDA antagonists), reversible cholinesterase inhibitors (such as huperzine), adenosine analogues, classical anti-epileptic treatment and cannabinoids are studied now such as promising possible new drugs to utilize to reduce seizures and to protect brain and vital organs [13].

It is important to improve and increase research programs in medical countermeasures, in developing protective equipment and carrying out training programs for medical and emergency personnel as well as for military nuclear, biological or chemical (NBC) units.

Acknowledgements

The authors express as sincere thank to the Italian Association of the Sovereign Military Order of Malta and to the San Raffaele Hospital Scientific Foundation of Milan, Italy for their precious support.

References

[1] E.T. Takafuji, A. B. Kok. *The Chemical Warfare Threat.* In: *Medical aspects of chemical and biological warfare.* Office of the Surgeon General, Department of the Army, U.S.A., Washington (1997)
[2] OECD *Environment Monograph* No. 81(OECD/GD 94(1): Paris, (OECD) (1994).
[3] J. Borak et al.. *Hazardous materials exposure: emergency response and patient care. New Jersey: Prentice Hall* (1991).
[4] A.C. Brostein, P.L. Currance. *Emergency Care for Hazardous Materials Exposure, Mosby Lifeline, 2nd ed.* (2004).
[5] NIOSH Pocket Guide to Chemical Hazards. Available at: http://www.cdc.gov/niosh/npg.html
[6] *Emergency Response Guidebook* 2004. Available at http://www.tc.gc/canutec/guide/guide.htm
[7] R. Harris, J. Paxman. *Paladin Books, London.* (1983)
[8] D. Baker. *PreHosp Disast Med* 19 (2) (2004)174.
[9] FR Sidell. *Nerve agents.* In: *Medical aspects of chemical and biological warfare.* Office of the Surgeon General, Department of the Army, U.S.A., Washington (1997)
[10] M.B. Abou-Dounia. *Chang Lw Dyer (Eds); Handbook of Neurotoxicology. Marcel Dekker, New York/Basel/Hong Kong:* (1995) 419.

[11] D. Grob. *Arch .Intern. Med* 98(1956):221.
[12] T. Zilker. *Toxicology* 214 (2005) 221.
[13] P. Aas. *Prehosp Disast Med* 18 (3) (2003) 208.
[14] V. Mirzanoyanov , L.Fedorov. *A poisoned policy. Moscow news.* 27 September 04 October: (1992)
[15] T. Stock . *Chemical and Biological weapons: Development and Proliferation. SIPRI Yearbook. World Armaments and Disarmaments, Oxford, UK,; Oxford University Press 1993*
[16] S. Mirzayanov. *Problems and Prospectis*. Report no. 17; Eds. Smithson AE, Mirzayanov S, Lajoie R, Krepon M. (1995)
[17] C Bismuth et al. *Toxicology Letters* 149 (2004)11.
[18] NATO handbook on the medical aspects of NBC defensive operations, AmedP-6(B), 1996.
[19] J.H. McDonough, T.M. Shih. *Neurosci Biobehav Rev* 17 (1993) 203.
[20] G. Lallement, F. Dorandeu F et al. *J Physiol (Paris);* 92(1998)369.

Chemical Warfare Agents: Medical Aspects and Treatment Principles, Part II

Alessandra ROSSODIVITA [a,1], Matteo GUIDOTTI[b], Massimo C. RANGHIERI[c], Elisaveta Jasna STIKOVA[d]

[a]*Department of Cardiothoracic and Vascular Diseases, San Raffaele Hospital Scientific Foundation; University "Life and Health", Milan, Italy*
[b]*CNR-Institute of Molecular Sciences and Technologies, Milan, Italy*
[c]*SMOM Auxiliary Corps of the Italian Army, 1° Reparto, Milan, Italy*
[d]*University "St Cyril and Methodius" Faculty of Medicine, Skopje, Republic of Macedonia*

Abstract: The most effective means of defending against chemical warfare agents, whether in war or as result of terror, are by use of primary and secondary prevention. The main goal of medical prevention programs is to minimize human loss by reducing the number of casualties, by teaching advanced medical programs in order to assure a proper response, and to develop new and effective countermeasures. The toxicity, medical management, characteristic diagnostic signs, and treatment of casualties with vesicant, blood, choking and incapacitating agents are described. The importance of improved research and learning processes for medical and emergency personnel as well as of military nuclear, biological and chemical (NBC) unit preparedness in the chemical agent field are emphasized.

Keywords: Chemical warfare agents, terrorism, toxicology, medical treatment, preparedness, chemical threat, vesicants, cyanide, choking agents, incapacitating agents

Since the 1995 Tokyo subway sarin attack, the threat of terrorist attacks involving weapons of mass destruction or industrial chemicals has produced worldwide insecurity and concern. Civilian exposure to toxic agents may be sudden or insidious. The most effective means of defending against chemical warfare agents, whether in war or as result of terror, are by use of primary and secondary prevention. The main goal of medical prevention programs is to minimize human loss by reducing the number of casualties, by teaching advanced medical programs that assure a proper response, and to develop new and effective countermeasures. Primary prevention involves preventing the use of, or exposure to, chemical agents. Since this is not

[1] Corresponding author San Raffaele Hospital Scientific Foundation, University " Life and Health", Department of Cardiothoracic and Vascular Diseases, 60 Olgettina Street, 20132 Milan, Italy; E-mail : rossodivita.alessandra@hsr.it

always possible, secondary prevention is also required, including proper medical and public health response, and new and effective medical countermeasures.

Military medical and nuclear-biological-and-chemical (NBC) units have dealt with mustard agent use and resultant casualties, such as in the recent Iran-Iraq war in the 1980's. The likelihood of military or terrorist use of other chemical agents is high.

This report describes the toxicity, medical management, characteristic diagnostic signs, and treatment of casualties with vesicant, blood, choking and incapacitating agents. This information should be part of the medical team's preparedness training for chemical attacks.

1. Vesicants

A *vesicant* is an agent that produces blisters or vesicles. Vesicants were first used as chemical weapons during World War I, causing about 80% of chemical casualties. Sulphur mustard remains a major chemical warfare agent. The vesicants are divided into three groups: 1) mustards, 2) phosgene oxime, and 3) arsenicals. Table 1 lists the naming system for mustard agents and vesicants according to the U.S. naming code [1].

1.1. Sulphur mustard

Sulphur mustard is the vesicant with the highest military significance. The last military use was in the Iran-Iraq War, where over 100,000 Iranians were injured by sulphur mustard and one-third still suffer from its late effects.

Mustard was named for its smell, taste (onion, garlic, mustard), and its color, which varies from yellow to dark brown. Mustard is an oily liquid and its volatility is low; thus it is "persistent". Sulphur mustard (di-2'-chloroethyylsulfide, $C_4H_8Cl_2S$, mustard agent) is an oily, clear, yellow-brown, liquid alkylating agent. It is poorly soluble in water. After aerosolization by bomb or shell blast, or by spraying, it vaporises slowly and may persist for one week in the environment in temperate climates. At higher temperature it vaporises more rapidly.

Sulphur mustard is a potentially effective terrorist or chemical warfare agent because it is widely available, stockpiled by over a dozen countries, easily manufactured, and inexpensive. The rapid and irreversible nature of its tissue reaction, combined with the delayed onset of symptoms, slows detection and worsens clinical effects. Sulphur mustard exerts a combination of biochemical effects on cells that result in its clinical manifestations. Key among these is alkylation of DNA, which activates chromosomal ADP-ribose polymerase, reduces the intracellular supply of oxidized nicotinamide adenosine dinucleotide (NAD+) thus inhibiting glycolysis. This in turn stimulates the hexose monophosphate shunt, which activates cellular proteinases, contributing to cell damage and dermal blister formation. Glutathione depletion is another possible mechanism of action, which results in compromised integrity of the cellular membrane structure and a cytokine-mediated inflammatory response [1, 2, 3].

Acute medical effects of exposure range from ocular and dermal injury, to respiratory tract damage, gastrointestinal effects and haematological effects. Chronic effects include reproductive and developmental toxicity, and cancer.

Table 1. Mustard and vesicant naming system according to the U.S. code [1]

U.S. code	Common Name
HD	Mustard
L	Lewisite (arsine)
CX	Phosgene oxime
T	Bis 2-chloroethylthioethyl
PD	Phenyldichloroarsine
ED	Ethyldichloroarsine

1.1.1. Impure Sulphur mustard (H)

Impure sulphur mustard smells like garlic or mustard. It appears as a pale yellow to dark brown liquid and it is persistent in soil. The median lethal dose is:

- LCt_{50}: 1,500 mg/min-m^3 (vapour)
- LD_{50}: approx 100 mg/kg (liquid)

1.1.2. Distilled Sulphur Mustard (HD)

Distilled sulphur mustard smells like garlic or mustard. It appears as a pale yellow to dark brown liquid and it is persistent in soil from 2 weeks to 3 years. The median lethal dose is:

- LCt_{50}: 1,500 mg/min-m^3 , vapour (inhaled); 10,000 (masked)
- LD_{50}: 100 mg/kg

Mustards are readily detoxified by diluted hypochlorite or soapy water [1].

1.1.3 Clinical effects

The first contact with sulphur mustard is mostly painless; the victim simply notices a garlic or sulphur odor. Normally, a symptom-free interval lasts for several hours. The duration of this interval correlates inversely with the adsorbed dose of the agent, and the maximum intensity of symptoms can be reached after days. The organs most commonly affected by mustards are those directly contacted, including the skin, eyes and airways. Exposure to large doses can cause damage to the hemopoietic and immune systems [4].

The onset of symptoms ranges from 1 to 12 hours after exposure. Effects on the skin range from erythema and edema to necrosis and vesicles. Although blisters generally form around 16 to 24 hours after exposure, they can form as late as 7 to 12 days later. Tracheobronchitis usually ensues several hours after exposure. Damage to the airways evolves with increasing (moderate to high dose) exposure, ranging from bronchospasm to bronchial obstruction to hemorrhagic pulmonary edema, to respiratory failure. Severe exposure causes oedema in the upper and lower airways with ulcerations and necrosis. Secondary bacterial pneumonia may complicate the course of treatment. The eyes are the most sensitive organ to sulphur mustard exposure. In comparison to skin, the latent period is shorter. Symptoms include erythema, edema, lachrymation, discomfort and pain. A high exposure results in severe eye pain, blepharospasm, iritis and blindness (either temporary or permanent). Nausea, vomiting and bone marrow suppression are additional clinical effects of high exposure. Sulphur mustard is called a "radiomimetic drug," as the effects of systemic poisoning are similar to those caused by exposure to ionizing radiation. Moderate dose exposure may cause headache, nausea and vomiting. High doses can severely damage the gastrointestinal tract and the bone marrow. This may result in immune suppression, leucopenia, diarrhea, fever, cachexia and seizures [5]. The long term effects include blindness, chronic bronchitis and pulmonary cancer.

1.1.4 Treatment

Until recently, there was no antidote for sulphur mustard. Treatment includes decontamination and supportive care, although sodium thiosulfate may prevent death, if given within minutes after exposure. Soldiers should immediately self-decontaminate; this is the only effective secondary prevention after exposure [1]. Decontamination of all skin surfaces with sodium hypochlorite 0.5% or copious water should occur as rapidly as possible, and before transport to the hospital. Skin lesions should be kept clean and monitored for infection; silver sulfadiazine or other topical antimicrobials may be useful. For patients with extensive skin injury it is crucial to monitor the hydration status. Pain can be treated with analgesics such as acetaminophen and opioids. Local or systemic corticosteroids may be required to control itching. Small vesicles should be not opened. Generally, sulphur mustard burns can be treated like second or third degree thermal burns. Severe respiratory distress may require ventilatory support. Bacterial infection may occur 3 to 8 days later. Early trachestomy is necessary in cases of stridor and hoarseness. Bronchoscopy may be necessary to remove pseudomembranes and other debris [1].

Eyes should be irrigated with water or saline 0.9% solution. Petroleum jelly (Vaseline) can be used to prevent the eyelids from sticking together and to maintain ocular fluid drainage [1]. Topical antibiotics and mydriatics can be used to prevent synechiae formation. Myelosuppression is the most serious effect of sulphur mustard exposure. Leucopenia develops on day 3 after exposure and reaches a nadir 7 to 9 days after exposure. A white cell count of less than 200 cells/mm^3 is a poor prognostic sign [6]. The delayed onset of clinical signs and symptoms after exposure to sulphur mustard is very typical. This delay may dangerously mask the need for treatment.

Medical personnel are advised to wear protective clothing, mask and gloves. In the military field, special skin decontaminants are used, including M291 (US army), RSDL™ (Canada), or Fuller's earth (UK). If these are not available, talcum powder or flour can be used [4, 7]. Late effects are fundamental to highlight and include respiratory effects, psychological disorders, skin symptoms and chronic eye lesions. Sulphur mustard is rated by the International Agency for Research on Cancer (IARC) as a human carcinogen and is a known risk factor for occupational lung cancer [8, 9].

1.2. Phosgene Oxime (CX)

Phosgene oxime is a colorless, crystalline solid. Its odor is intense and irritating. It persists in the soil for 2 hours. Median lethal doses are:

- LCt_{50}: 3,200 mg/min-m^3 (estimate)
- LD_{50}: no estimate

Not a true vesicant, phosgene oxime does not produce vesicles. Rather, it is an urticant, causing erythema, wheals and urticaria. It causes extensive tissue damage, and therefore is labelled as a corrosive. It differs from phosgene (CG), which is a lung agent that attacks the alveolar-capillary membrane [1].

CX affects the skin, eyes and the lungs. The effects are quicker and more damaging than other vesicants – it causes immediate pain and irritation of the skin, eyes and airways.

There is no antidote for phosgene oxime. Medical personnel should treat the necrotic lesions in the same way they treat other necrotic lesions. Skin lesions should be kept clean and monitored for infection; silver sulfadiazine or other topical antimicrobials may be useful. Eyes should be irrigated with water and injuries should be treated with topical antibiotics. Severe respiratory distress might require ventilatory support, and severe bronchospasm may require beta-agonists and corticosteroids [1].

1.3. Arsenical – Lewisite (L)

Lewisite is a colorless oily liquid with a typical geranium odor. It is persistent in soil for days, and is more volatile than mustard. The median lethal dose is:

- LCt_{50}: 1200 – 1500 mg/min-m^3 (inhaled); 100,000 (masked)
- LD_{50}: 40 – 50 mg/kg

Lewisite is an organic arsenical compound. Its symptoms are immediate. It primarily inhibits thiol-containing enzymes, disrupting energy pathways, depleting adenosine triphosphate (ATP), causing cell death [1].

1.3.1. Clinical effects

Clinical effects are similar to those caused by mustard except that the onset of symptoms is earlier and pain and erythema are immediate. With high exposures there can be a potentially lethal "Lewisite shock" due to an acute depletion of intravascular volume secondary to capillary leak. High dose exposure can also result in hepatic necrosis and renal failure.

1.3.2. Treatment

Lewisite is detoxified by dilute hypochlorite, soapy water, or adsorbent powders. Decontamination should be performed immediately, with all clothing removed. The eyes should be flushed and irrigated with large amounts of water. Supportive care of skin, eye and airway injury is similar to that for mustard exposure. British Anti-Lewisite (BAL, 2,3-dimercaptopropanol), which chelates Lewisite and reactivates enzymes, should be administered at a dosage of 3 to 5 mg /kg IM every 4 h for 4 doses to patients who are in shock or have severe pulmonary involvement. Topical application of BAL to skin and eyes (concentration > 20%) is indicated to prevent injury and vesication. Contraindications include pregnancy and pre-existing renal failure. Alkalinisation of the urine may help to prevent BAL-induced renal damage. Systemic and topical analogues of BAL are under investigation [1].

2. Blood agent – cyanide

Cyanide compounds are the prototype members of the group of chemical agents that are termed blood agents. They attack and disable the oxygen-carrying capacity of blood.

Cyanide has been used as a poison for thousands of years. The North American Treaty Organization (NATO) specifies cyanides of military interest as hydrogen cyanide (AC) and cyanogen chloride (CK) [10]. Other related cyanide compounds such as cyanide salts are less effective combat poisons. Cyanides are less toxic than many other chemical warfare agents. Cyanide binds and inactivates several enzymes. Its cellular effects are thought to occur by binding the active sites of cytochrome oxidase enzymes, stopping aerobic cell metabolism after an initial effect on excitable tissue.

Effects of cyanide poisoning include progressive histotoxic tissue hypoxia. Symptoms, signs and physical findings are directly related to the dose, the route of exposure, and the type of cyanide compound. Symptoms of mild poisoning include headache, dizziness, drowsiness, nausea, vomiting and mucosal irritation. Severe symptoms include dyspnea, impaired consciousness and coma, convulsions, tachy- and brady-dysrhythmias, hypotension, cardiovascular collapse and death, which can occur within minutes of inhalation. Metabolic acidosis results from lactic acid accumulation and there is usually a decrease in the arterial-venous difference in the partial pressure of oxygen. Cyanide poisoning should be suspected in any acyanotic patients who appear to suffer from severe hypoxia. The smell of bitter almonds on the breath or in gastric washings can be a clue to diagnosis. Whole blood cyanide levels often require several days for analysis at a specialized hospital toxicology center [11, 12].

2.1. Principles of Therapy

It is important to eliminate further exposure and institute supportive treatment and specific antidotes. Supportive treatment includes administration of high concentrations of oxygen, correction of acidosis, treatment of seizures and cardiovascular support. Casualties with advanced toxicity from high-dose cyanide exposure may also require specific antidote therapy. Cyanide is normally detoxified by rhodanase (thiosulfate-cyanide sulphur transferase) which binds sulphur to cyanide, to form thiocyanate, which is excreted in the urine.

Sodium nitrite and sodium thiosulfate (IV), dicobalt edetate (IV), and hydroxocobalamin are the most recommended antidotal treatments; they vary according to country and medical custom. The solution of sodium nitrite (Lilly Cyanide antidote kit – 30 mg/mL) should be given to an adult intravenously over 5 to 15 minutes, with careful monitoring of blood pressure. The recommended dose for children is 0.33% mL of the 10% solution per kilogram of body weight. Sodium thiosulfate (Lilly kit) is administered as 50 ml into 250 mg/mL of saline intravenously. Hydroxocobalamin (vitamin B_{12}) is recommended in a dosage of 4 g to neutralize a lethal dose of cyanide [1, 2, 10, 12, 13].

2.2. Hydrogen cyanide (AC)

Hydrogen cyanide is a colorless gas or liquid that smells like bitter almonds. It is highly volatile, which limits its use in confined spaces. The median lethal dose is:

- LCt_{50}: 2500 – 5000 mg/min-m^3 (time dependent)
- $LD_{50\,(skin)}$: 100 mg/kg

2.3. Cyanogen Chloride (CK)

Cyanogen Chloride is a colorless gas or liquid; non persistent. Median lethal dose is:

- LCt_{50}: 11,000 mg/min-m^3
- LD_{50}: not listed

Cyanides are detoxified by water or soap and water [1, 2, 10].

3. Lung damaging agents (choking agents).

Chemical agents that attack lung tissue, primarily causing pulmonary oedema, are classified as lung-damaging agents. This group includes phosgene (CG), diphosgene (DP), chlorine (Cl) and chloropicrin (PS). Certain other substances, while not likely to be used as agents, are still likely to be met on the battlefield (e.g., nitrous fumes and zinc chloride smoke) and may have a similar action. Similar substances encountered in fires, e.g., perfluoroisobutylene (PFIB) and HCl, may also induce lung damage [14, 15].

3.1. Phosgene (CG)

The toxic action of phosgene is typical of lung-damaging agents. Phosgene is the most dangerous member of this group and the most likely to be used in the future. It was first used in 1915, and accounted for 80% of all chemical fatalities during World War I. Phosgene causes extensive lung parenchymal damage. It causes injury through two different chemical reactions, acylation and hydrolysis. Acylation denatures proteins, causing cell membrane and enzyme dysfunction. Hydrolysis of phosgene to hydrochloric acid injures the eyes and the upper and lower respiratory tract in moderate to high concentrations. The main lesion of phosgene in the lungs is high permeability pulmonary edema that may be accompanied by necrotizing bronchiolitis and thrombosis of pulmonary venules. It can evolve into acute respiratory distress syndrome (ARDS).

There is no specific antidote for phosgene. Treatment is supportive. Patients exposed to phosgene should be monitored closely for symptoms and signs of pulmonary edema, with chest X-rays and blood gas testing. ARDS may require ventilatory support. Beta-agonists and corticosteroid treatment are still controversial [14, 15, 16].

3.2. Chlorine (Cl)

Chlorine is a pungent yellow-green gas with intermediate water solubility. It injures the upper airway, the conducting airways, and the alveoli themselves. Chlorine reacts with water to form hydrochloric and hydrochlorous acids. These acids generate oxygen free radicals that can penetrate cell membranes. They form chloramines that oxidise sulphur-containing amino-acids, leading to cell injury and death. In addition to ocular and upper airways injury, chlorine can cause both acute and chronic lung diseases [12, 13, 14], although most acutely exposed victims recover fully [16]. Severe acute lung injury with ensuing ARDS is possible.

Diagnosis and treatment of exposed victims is similar to that for phosgene exposure, and also includes bronchodilator (e.g., beta-agonist) and corticosteroid treatment for bronchospasm, as necessary [16].

4. Incapacitating agents

Incapacitating agents are non-lethal chemical agents. The term "incapacitate" means to "deprive of strength or ability". In a military context, incapacitation means the "inability to perform one's military mission" [17]. An incapacitant is a chemical agent that produces a temporary disability that can persist for hours or days after exposure. Medical treatment, while not essential, may facilitate a more rapid recovery. An incapacitating substance has a severe and rapid but non-lethal effect, altering the CNS. They are not likely to produce permanent injury in concentrations which are militarily effective. Any agent that can disrupt aspects of vital performance or any military task, producing a temporary disability, could be considered an incapacitating agent. The group includes drugs with psychochemical or behavioral effects, which fall into four categories: 1) stimulants, 2) depressants, 3) psychedelics, and 4) deliriants [17].

1) **Stimulants**: amphetamines, cocaine, caffeine, nicotine, and epileptogenics
2) **Depressants:** barbiturates, morphine, opioids, haloperidol, buthyrophenones
3) **Psychedelics or CNS stimulants:** LSD; MDMA (3,4-methylenedioxymethylamphetamine, "ecstasy"); PCP (phencyclidine), ketamine
4) **Deliriants:** anticholinergics – e.g., BZ

On a battlefield, or for terrorists use to cause mass injury, the most effective agents of this group are the anticholinergics. These drugs block the muscarinic effect of acethylcoline in the CNS and the peripheral nervous system. The best known are atropine and scopolamine [17]. The anticholinergic BZ (3-quinuclidinyl benzylate) is one of most powerful anticholinergic substances. Its clinical profile closely resembles atropine, differing in duration and potency and causing delirium at low dosage. The signs and symptoms of attack by anticholinergics include dry mouth, tachycardia, elevated temperature, blurred vision, papillary dilatation, restlessness, dizziness, confusion, slurred or nonsense speech, failure to obey orders, hallucinations, mumbling, stupor and coma [15, 17].

4.1. Medical treatments

1) BZ and other anticholinergics agents – physostigmine IV 30 mcg/kg or 4 mg IM, or 2 mg orally every 2 h;
2) LSD – chlorpromazine (50 to 100 mcg/kg PO); or lorazepam (10 - 20 mg IV);
3) Opioids – naloxone 0.4 - 1.0 mg sc or IV [15, 17].

5. Riot control agents

Riot control agents are compounds that temporarily incapacitate victims by irritation. They cause profuse eye symptoms, oropharyngeal and nasopharyngeal pain, rhinorrhea, and upper airway irritation. The agents are called "tear gases" and have common characteristics. These include rapid onset of effects and a relatively brief duration (25 to 30 minutes). They are often called irritating agents. Three categories of riot agents are recognized: 1) lachrymators, 2) sternutators (which cause sneezing and upper respiratory irritation), and 3) vomiting agents. The agents most commonly used are chloroacetophenone (CN) and orthochlorobenzylidene-malononitrile (CS). Heavy exposure inside closed structures results in acute lung injury, reactive airways dysfunction syndrome, and death [2, 17]. Table 2 lists the principal riot control agents.

Table 2. Naming system for riot agents [17]

U.S.	Common Name
DM	Adamasite
Chloropicrin	Vomiting Gas
CN	Police Tear Gas
CR	British Tear Gas
CS	Army riot control
CA	Similar to CN
CNC	Mace

5.1. 1-o-Chlorobenzylidene malononitrile (CS)

CS is a white crystalline powder with a pungent odor like pepper. Its effects appear within seconds of exposure to an aerosolized compound. Most of the effects only last for about 30 minutes, although mild erythema can persist for 1-2 h. Clinical effects are localized to eyes, skin and airways. In the eyes, lachrymation, blepharospasm and conjunctival injection can occur. Photophobia is often present and may persist for one hour after the exposure. The mucous membranes of the mouth, tongue and palate have a sensation of discomfort or burning with excess of salivation. In some cases vomiting and marked coughing occur [14, 15, 17].

5.2. 1-Chloroacetophenone (CN)

CN smells like apple blossoms. It is a solid powder and can be disseminated as a smoke generated from a grenade or other device, or in powder or liquid formulations. The clinical effects are the same as those caused by CS. Most effects from exposures to low concentrations disappear within 20- 30 minutes. Use of large amounts of CN in confined spaces has caused injuries requiring medical attention and, rarely, death [14, 15, 17].

5.3. Diphenylaminearsine (DM)

DM is one of the groups that is known as vomiting agents. It is a yellow-green, odorless crystalline substance with low volatility. It primarily acts on the upper respiratory tract, causing irritation of the nasal mucosa and nasal sinuses, burning in the throat, tightness and pain in the chest, and uncontrollable coughing and sneezing. It also causes eye irritation and burning. This compound has two main characteristics: its effects do not appear immediately on exposure; and it may cause prolonged systemic effects such as headache, mental depression, chills, nausea, abdominal cramps, vomiting and diarrhea, which last for several hours after the exposure [14, 15, 17].

5.4. Dibenz(b,f)-1:4 oxazepine (CR)

CR is a relatively new compound. CR is more potent and less toxic than CS. The primary effects are eye pain, blepharospasm and lachrymation which persist for 15 to 30 minutes. A transient erythema can occur in 1 to 2 h.

The effects of riot agents are usually self-limiting and do not require medical attention. In rare circumstances complications of the skin, eyes and airways can occur. All of these agents are detoxified with soap and water [14, 15, 17]. The use of water can increase a transient worsening of symptoms [14, 15, 17].

6. Acknowledgements

The authors express sincere thanks to the Italian Association of the Sovereign Military Order of Malta and to the San Raffaele Hospital Scientific Foundation of Milan, Italy for the precious support.

7. References

[1] FR Sidell, JS Urbanetti, WJ Smith, CG Hurst. *Vesicants.* In: *Medical aspects of chemical and biological warfare.* Office of the Surgeon General, Department of the Army, U.S.A., Washington (1997)
[2] C Bismuth, S W Borron, FJ Baud, P Barriot. *Toxicology Lett.* 149 (2004) 11
[3] RA Greenfield, BR Brown, J B Hutchins, JJ Iandolo, R Jackson, LN Slater, M. S. Bronze, *Am J Med Sci,* 323(6) (2002) 326.
[4] K Kehe, L Szinicz. *Toxicology* 214 (2005) 198
[5] JC Dacre, M Goldman. 1996. *Pharmacol Rev* 48 (2) (1996)289
[6] JL Williams. *Ann Med Mil* 3 (1) (1989)15
[7] J Borak, FR Sidell. 1992. *Ann Emer Med* 21(1992)303
[8] S Wada, M Miyanishi, Y Nishimoto, S Kambe, RW Miller. *Lancet* 1(1968) 1161
[9] Y Nishimoto, et al. 1986. *Gan To Kagaku. Ryoho* 13 (1986) 1144
[10] SI Baskin, TG Brewer. *Cyanide poisoning.* In: *Medical aspects of chemical and biological warfare.* Office of the Surgeon General, Department of the Army, U.S.A., Washington (1997)
[11] AH Hall, BH Rumak. *Ann Emerg. Med* 15(1986)1067
[12] T Zilker. *Toxicology* 214 (2005) 221.
[13] D Baker. *PreHosp Disast Med* 19 (2) (2004)174.
[14] NATO handbook on the medical aspects of NBC defensive operations, AmedP-6(B), 1996.
[15] World Health Organization, *Public health response to biological and chemical weapons: WHO guidance.* Geneva, 2004.
[16] JS Urbanetti. *Toxic inhalants.* In: *Medical aspects of chemical and biological warfare.* Office of the Surgeon General, Department of the Army, U.S.A., Washington (1997)
[17] JS Ketcham, FR Sidell. *Incapacitating agents.* In: *Medical aspects of chemical and biological warfare.* Office of the Surgeon General, Department of the Army, U.S.A., Washington (1997)

Strengthening National Public Health Preparedness and Response to Chemical, Biological and Radiological Agent Threats. Edited by C.E. Cummings and E. Stikova.
IOS Press, 2007.

Chemical Warfare Agents: Weapons of Mass Destruction or Psychological Threats?

Matteo GUIDOTTI [a,b,1], Massimo C. RANGHIERI [b] and Alessandra ROSSODIVITA [c]
[a] *CNR-Institute of Molecular Sciences and Technologies, Milan, Italy*
[b] *SMOM Auxiliary Corps of the Italian Army, 1° Reparto, Milan, Italy*
[c] *Dep. of Cardiothoracic and Vascular Diseases, San Raffaele Hospital, Milan, Italy*

Abstract. In the last century, the deliberate use of chemical warfare has inflicted mass casualties and remarkable psychological consequences on the public. A historical summary of the most relevant chemical warfare agents is reported, as well as the main advantages and disadvantages to the use of chemical weapons in warfare or terrorist attacks in the beginning of the 21st century. Toxic aggressives are described according to their physiopathologic effects on the human body. Some prevalent key myths about chemical weapons are discussed and discredited. Even though the risk of use of chemical agents for warfare purposes has diminished over the past years, these weapons still represent a potential threat in episodes of national and international terrorism. Accurate information and basic training campaigns for civilian populations can reduce anxiety levels, improve the general preparedness, and limit the terrorists' capacity for harm.

Keywords. Chemical weapons, warfare agents, preparedness, terrorist attack, chemical threat, NBC defense

Introduction

In the 1980s and '90s, the agreement on the non-proliferation of chemical weapons (CW) and stockpile disposal programs optimistically suggested that these highly toxic compounds would rapidly disappear within the span of several decades. Unfortunately, this scenario changed abruptly in recent years as unstable international events forced many countries around the world to pay renewed attention to the potential use of chemical weapons for warfare and terrorist attacks. Concerns about the uncontrolled diffusion of chemical weapons are driven by questionable government ethics and scientists' complicity. In addition, the ever-growing media interest in these topics often spreads inaccurate information, increasing civilian anxiety. The lack of a thorough knowledge about the real hazards linked to the use, handling and decontamination of these compounds may lead to a bad evaluation, or even to an overestimation of the risks and the consequences of chemical agent attacks. However, the physico-chemical characteristics of the agents used as CW are sometimes similar to those of several highly toxic chemicals employed in pacific industrial applications (*e.g.* organophos-

[1] CNR-Institute of Molecular Sciences and Technologies, via G. Venezian 21, 20133 Milano, Italy; e-mail: m.guidotti@istm.cnr.it.

phates for insecticide production or cyanides for organic chemical industry). Therefore, many experts, including chemists, toxicologists or physicians, are competent in managing hazardous materials and in minimizing the effects of potential accidents.

For these reasons, it is necessary to disseminate correct information regarding chemical warfare agents at different levels to the public, as well as to emergency responders. Multidisciplinary divulgation and training can bridge the gap between academics, which often consider CW as "taboo" poisons that do not deserve any sort of interest, and technical professionals (military and civilian defense authorities), which play a primary role in maxi-emergencies and which already have an adequate knowledge about CW. A careful divulgation should remove some people's misconceptions regarding these agents and address the basic principle that timely and accurate countermeasures are extremely effective in minimizing the consequences of a CW use. This training can dispel common mistaken beliefs such as the idea that CW agents are invincible and almighty. A result could be a reduced psychological threat, and a diminished probability that CW agents will be used.

1. Historical Overview

The use of toxic and irritant chemicals in warfare has been reported since antiquity. The Latin sentence "*armis bella non venenis geri*" (wars are fought with weapons, not with poisons) marked the first condemnation of these non-conventional weapons in the 1st century A.D. by Roman jurists. However, the first systematic and massive use of chemical weapons was recorded during World War I. The large number of victims caused by these new warfare agents led to the development of the Geneva Protocol in 1925. One hundred and thirty countries signed this protocol, proving the international community acknowledged the highly destructive potential of chemical weapons. Within this context, chemical weapons were defined as "those chemical substances, whether gaseous, liquid or solid, which might be employed because of their direct toxic effect on man, animals and plants" [1]. The right to use CW in retaliation against an attack with such agents was reserved, however. Therefore, the treaty was in effect a no-first-use agreement. States parties of the protocol eventually infringed the treaty and employed chemical agents in warfare (as, in 1930s, Italy in Ethiopia or Japan in Manchuria).

During World War II, despite the very large stockpile of CW agents, no deliberate use of CW was recorded. After a period of static deterrence, the employment of CW in the local wars was questionable, as evidence suggests that these agents were used during the Iran-Iraq conflict (1981–1989). Furthermore, the development of long-range missiles dramatically increased the threat of chemical warfare, since missiles can carry the lethal effect of these agents to far-reaching areas.

The renewed threat and advances in scientific capabilities of weapons production [2,3] led to the signing of the Chemical Weapons Convention by 111 countries in Paris, January 1993. The Convention defines chemical weapons as not only the toxic chemicals, but also the ammunition and equipment for their dispersal. According to this treaty, CW are defined as "any chemical which, through its chemical effect on living processes, may cause death, temporary loss of performance or permanent injury to people and animals" [4]. Unlike the Geneva Protocol, the Convention attempts to list the prohibited substances (the toxic compounds as well as the chemical precursors needed to synthesize CW) and to establish enforcement mechanisms. In addition to banning

CW use, the Convention bans the development, production, stockpiling and transfer of the chemical agents. However, it does not prohibit the study of toxic compounds for industrial, agricultural, medical or pharmaceutical purposes.

2. Some Commonplaces to Discredit

It is important to discredit several key myths about chemical warfare agents. For instance, these substances are frequently called *war gases* or *toxic gases*. These incorrect terms are a heritage of history. During World War I, gaseous or low-boiling liquids, such as chlorine or phosgene, were employed. At the present time, gases as CW are rarely used, and high-boiling liquids or solids are preferred. If volatile agents must be used, thickening inert components are often added to increase the persistence in the environment and the offensive efficacy.

Another such myth is that military research has recently developed new and more dangerous chemical agents. In last 40 years, innovation in the field of CW research has been minimal, compared with other fields of weaponry (mainly nuclear and biological weapons). Most chemical agents were developed between the 1930s and the 1960s [5]. In fact, even though more than 100,000 toxic compounds have been examined for potential use as warfare agents, only around 70 chemical species were developed and produced in large amounts for the arsenals of many countries [6,7]. Rather, considerable efforts were devoted to developing new detection and defense techniques, improving decontamination methodologies, and destroying the stockpiled CW in an environmentally sustainable way [8–10]. This trend was also due to more stringent international restrictions on the disposal of CW and of their production facilities [11,12].

Another common myth is that CW are extremely lethal. While extremely toxic, CW are only 5% fatal on battlefields due to the complications in their dissemination and dispersal (for chlorine in WWI and for mustard agents in Iran-Iraq war a mortality of 2% and 4%, respectively) [13]. However, CW agents inflict large numbers of injuries, causing a significant logistical burden to opponent forces.

Obviously, these considerations do not apply to toxins, the highly toxic compounds produced by living organisms (or their synthetic equivalents) classified as chemical warfare agents if they are used for military purposes. Recent advances in genetic engineering and in biotechnologies have allowed for a wider library of available toxins and larger production capabilities [14–17]. However, these weapons of biological origin have a special role in the array of measures against CW proliferation, since they are covered by the Biological and Toxin Weapons Convention of 1972 [18]. This convention bans the development, production and stockpiling of such substances not required for peaceful purposes. For the purposes of this report these toxins will not be included.

3. Advantages and Disadvantages

The practical use of CW presents a series of different characteristics that, according to situations can be considered as advantages or disadvantages.

Chemical weapons:

- are relatively inexpensive and, compared with other mass destruction weapons, are simple to produce;
- are obtained by easily accessible and well-known technologies;

- require large amounts of chemical precursors to set up an efficient arsenal;
- affect opposing forces without damaging structures and materials;
- have psychologically devastating effects on civilian populations;
- can be almost completely neutralized by timely defensive measures;
- are highly susceptible to environmental conditions (moisture, temperature, wind, etc.);
- reduce the combat efficiency of the troops, both in offensive and in defensive use;
- require suitable dispersion and dissemination devices for an effective use;
- require large (and safe) facilities for their storage and stockpiling.

It is evident that the use of CW in modern warfare is marked by more disadvantages than advantages, especially if these substances are employed against well-trained, well-equipped troops. Several factors have contributed to a recent and gradual decline in the production of CW. These include that weaponization of most chemical agents is difficult, the threat that the use of CW might lead to nuclear retaliation, and the consideration that conventional explosives inflict, on average, seven times more damage than CW at equal charge and range [14]. On the contrary, in the case of a terrorist attack, the use of CW against unprepared civilian populations could lead to catastrophic consequences in terms of casualties and of psychological effects on the public [19]. This was the case in 1995 when the fanatic religious sect, Aum Shinrikyo, launched a multicentric attack in the Tokyo underground: some perforated bags containing the nerve agent sarin killed 12 people and injured approximately 500 [20,21].

4. Classification of the CW

Chemical warfare agents are typically categorized according to their physiopathologic effects on the human body [18,22–25]. First, they are grouped in to two main categories: *lethal aggressives* and *incapacitating aggressives*. An agent is incapacitating if 1/100 of its lethal dose is able to cause a psychological or physiological reaction and severely impair an individual's ability to perform a mission. The boundary between lethal and incapacitating agents is not absolute, but it is based on toxicological and statistic evaluations. Lethal aggressives are classified as: neurotoxic ('nerve') agents; vesicant agents; blood agents; choking agents. Incapacitating aggressives are classified as: psychochemical agents and irritants (vomiting or tear agents). Table 1 shows the main physico-chemical characteristics of the most widely-known CW.

4.1. Neurotoxic Agents

Since World War II, nerve agents have played a key role as stockpiled weapons. Nerve agents are the most toxic of the CW agents [13,14,27] and belong to the family of organophosphorus compounds. Most of them are alkyl esters of cyanophosphoric acid (*e.g.*, tabun) or of methylphosphonic acid (*e.g.*, sarin or soman). Nerve agents acquired their name because they inhibit the enzyme acetylcholinesterase, which controls the transmission of nerve impulses, leading to an abnormal excess of acetylcholine in the cholinergic synapses. Under normal conditions, the enzyme rapidly breaks down by

Table 1. Physicochemical and toxicological properties of some CW (adapted from [22–26])

1	**Sarin**	**VX**	**Hydrogen Cyanide**	**Phosgene**	**Chloropicrin**	**Sulfur Mustard**
2	nerve	nerve	blood	Choking	choking	vesicant
3	–56	–39	–14	–118	–69	14
4	147	298	26	8	112	228
5	$1.6 \cdot 10^4$	3	$8.7 \cdot 10^5$	$6.4 \cdot 10^6$	$1.7 \cdot 10^5$	625
6	complete	1–5%	complete	hydrolyzes	0.2%	0.1%
7	5	0.5	$2.0 \cdot 10^3$	$1.6 \cdot 10^3$	60	100
8	70–100	50	1000–2000	5000	8000	1000
9	1700	6	7000	–	–	8000

1	**Lewisite**	**LSD**	**BZ**	**Adamsite**	**CS**
2	vesicant	psychochemical	psychochemical	Irritant	irritant
3	–13	198	240	195	72
4	190	–	–	410 (dec)	310
5	$4.5 \cdot 10^3$	negligible	negligible	$2.9 \cdot 10^{-2}$	0.35
6	low	high	high	0.6%	0.05%
7	300	–	100–200	20–25	5–10
8	1200	–	$2 \cdot 10^5$	$1.5 – 3.0 \cdot 10^4$	$0.3 – 1.0 \cdot 10^5$
9	2500	–	–	–	–

Legend: (1) common name; (2) class; (3) melting point (°C); (4) boiling point (°C); volatility at 20°C(mg m^{-3}); (6) solubility at 20°C; (7) ICt_{50}: median incapacitating exposure (mg min m^{-3}); (8) LCt_{50}: median lethal exposure (mg min m^{-3}); (9) LD_{50}: lethal dose (dermal) (mg person^{-1}).

hydrolysis of acetylcholine, which mediates the nerve signal and brings the receptor back to the original rest situation (Fig. 1a).

When a nerve agent is present, the enzymatic activity is suppressed and the neuronal receptor is "blocked" by the large overload of acetylcholine that cannot be hydrolyzed (Fig. 1b). Consequently, the transmission of the nerve impulse sequences is completely jammed. The poisoning is extremely rapid in onset and severe symptoms quickly lead to death. From a molecular point of view, the reactive center is the phosphorus atom that easily reacts with the hydroxyl functional group of the serine present at the active site of the enzyme, which is therefore irreversibly deactivated.

Similar organophosphorous compounds are also employed in the manufacturing of insecticides and flame retardants. The remarkable similarity between nerve agents and their industrial counterparts shows how simple it would be for small groups of terrorists to obtain or purchase toxic chemical precursors for the small-scale production of neurotoxic CW.

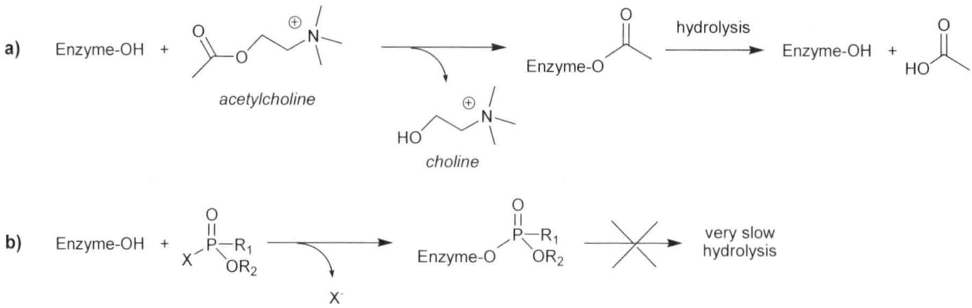

Figure 1. Acetylcholinesterase function under normal conditions (a) and after exposure to nerve agents (b).

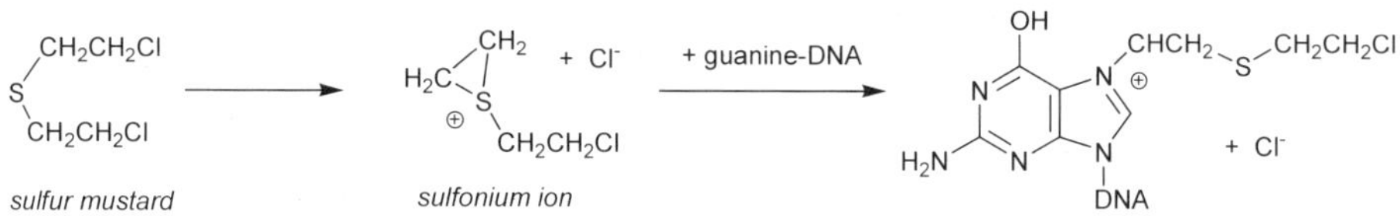

Figure 2. Reaction of sulfur mustard with DNA.

4.2. Vesicant Agents

Vesicants agents or mustard agents were available since the 1820s, but were not used in warfare until WWI. They are called mustard agents because they have a characteristic and sharp smell similar to mustard when obtained in low purity. From a chemical point of view, these agents are chloroalkyl derivatives of sulfur or nitrogen. The outstanding affinity of these substances to covalently bind the nitrogen atoms of the nucleic acids (typically they react with the N7 of the guanine; Fig. 2) or the -SH functional groups of proteins and enzymes accounts for their high toxicity [11]. Since each molecule possesses many reactive centers (one for each chlorine leaving group), it can form intra- or intermolecular bridges and therefore hinder, or even block, numerous metabolic processes in living tissues. Symptoms associated with vesicants poisoning include appearance of burns and blisters in the skin and eyes. Vesicants are particularly insidious; symptom onset can range from two to 24 hours after exposure.

Several synthetic routes can be followed to easily produce vesicant agents, none of which requires sophisticated technologies or unusual precursors. In addition, because of their low cost, their predictable properties and their ability to cause resource-devouring casualties rather than fatalities, mustard agents are often the first choice of countries who want to build their capability for chemical warfare.

4.3. Blood Agents

Blood agents include compounds such as hydrogen cyanide (HCN) or carbon monoxide (CO), whose toxicity and poisoning activity have long been identified. Blood agents cause poisoning by ingestion, skin absorption or inhalation. These agents are all simple molecules, with a low molecular mass and a remarkable affinity to bind the transition metals, especially iron. The most important toxic effects of blood agents are the result of inhibition of metal-containing enzymes, such as cytochromoxidase (containing iron), which is responsible for the aerobic processes of cell respiration. If the metal active site is irreversibly blocked, the aerobic energy production is stopped and the cell dies rapidly.

Although these agents are very toxic, their extreme volatility (they are really *war gases*) limits their use outdoors in warfare. Conversely, since they can be highly lethal in confined spaces (buildings, bunkers, armored vehicles, etc.), they could remain potential choices for terrorist attacks and pose a serious threat.

4.4. Choking Agents

Chlorine (Cl_2) and phosgene ($COCl_2$) are important chemicals for the synthetic industry and they were the first CW used during WWI. Large amounts of agents were required to produce militarily significant effects (generally only against unprepared and unsophisticated opponents), but they were very effective in spreading the panic and the fear of the *invisible death* among the troops. Choking agents are gases or highly volatile liquids with pungent odor. When inhaled, they irritate the eyes and the respiratory tract. Even at low concentrations, they can inflict severe lung damage. Biochemically, these substances cause injury via both hydrolysis and binding to proteins: the hydrochloric acid formed in vivo causes injury to eyes and upper respiratory tract, whereas the formation of free radicals and/or acylating agents denatures proteins, leading mainly to cell membrane and enzyme dysfunction.

As are most of the blood agents, phosgene and chlorine are commonly used worldwide in industrial manufacturing (synthesis of chemicals, pharmaceuticals, agrochemicals, etc.). The threat posed by the military use of these agents is lowered by the use of protective equipment and training. However, attacks with these simple CW agents on unprotected civilian populations in densely populated areas could result in numerous casualties.

4.5. Psychochemical Agents

During the Cold War in the second half of the 20th century, military research considered the feasibility of developing nonlethal weapons. Although they were not a priority in the U.S. at the time, funding for related research has increased [28]. Nonlethal weapons are drugs that induce a temporary psycho-neurological incapacitation and impair completely the fighting ability of a combatant. Among the best known and most potent psychoactive agents are N, N-diethyl-D-lysergamide (better known as LSD) and 3-quinuclidinyl benzilate (BZ) (Fig. 3). Their action mechanism is based on alteration in the transmission of nerve impulses from the peripheral to the central nervous system and on the modification of sensory perceptions. Such alteration can induce depression (as with BZ) or delirium (as with LSD) in the exposed individuals and lead to psychotic and behavioral disorders [28].

The production of these agents is similar to the manufacturing of pharmaceuticals. In fact, most of the psychochemical agents have legitimate therapeutic uses. However, the need for large-scale industrial production facilities (larger than for conventional drug production), high manufacturing costs, and the difficult incorporation of these agents in weapons are the main deterrents to their use as CW.

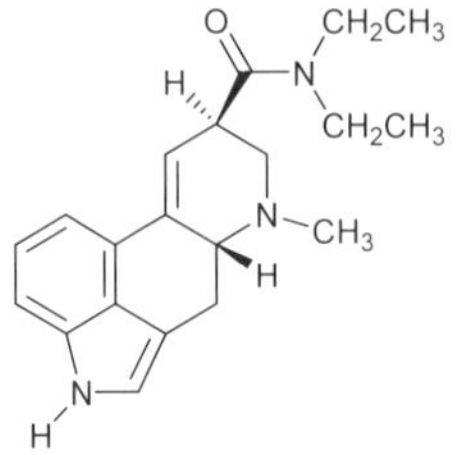

N,N-diethyl-D-lysergamide (LSD)

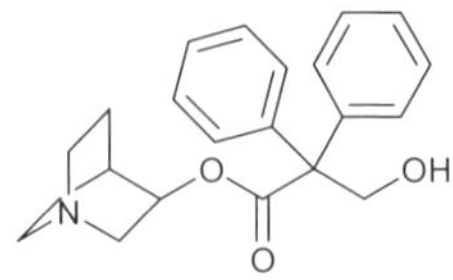

3-quinuclidinyl benzilate (BZ)

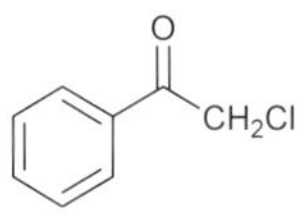

chloroacetophenone (CN)

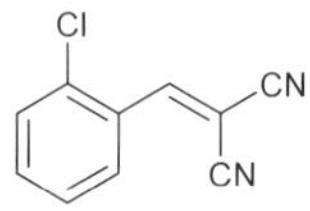

ortho-chlorobenzylidene malononitrile (CS)

Figure 3. Incapacitating agents.

4.6. Irritating Agents

The use of irritating smokes, obtained by burning mixtures of pitch, sulfur and other compounds, has been recorded since antiquity, but were only systematically studied after WWI. The irritants are classified into sub-groups according to the most prominent effect on the human body: *lacrimators* (*tear gases*), *sternutators* and *vomiting agents,* which cause instant irritation in the eyes, nose and throat, respectively. All of the irritating agents have a very high safety ratio, *i.e.* a large gap between the threshold concentration (the concentration required to obtain a perceptible effect on the 50% of the exposed subjects) and the mean lethal concentration (the concentration causing the death to the 50% of the exposed subjects). Symptoms have a rapid onset (seconds to minutes) and only last for about 15–30 minutes post-exposure. For these reasons, tear agents are widely used by police forces for law enforcement. Chloroacetophenone (CN) and, most of all, *ortho*-chlorobenzylidene malononitrile (CS) are the standard irritating agents for military use and for civil riot control (Fig. 3).

The use of these agents in warfare (between two sovereign states) is prohibited by the Chemical Weapons Convention, but the treaty does not prohibit the development of toxic chemicals for law enforcement purposes, including domestic riot control.

Conclusions

Although the perceived risk of use of CW in warfare has always had a high psychological impact on the public, the real threat has gradually diminished in recent years. This trend has been confirmed by the relatively slow development of chemical agent technologies, compared with nuclear and biological weapons. Training and protection is essential to reduce the efficiency of CW in a war scenario. CW, however, still represent a threat in episodes of terrorism or sabotage by small sub-national independent or state-sponsored groups. In this case, accurate information and basic training campaigns for civilian populations can reduce anxiety levels, improve the general preparedness, and limit the terrorists' capacity for harm.

Acknowledgements

M. Guidotti and M.C. Ranghieri wish to thank Prof. P. Rumm, for the kind invitation to the NATO Advanced Study Institute held in Skopje (Macedonia), and the Italian Association of the Sovereign Military Order of Malta.

References

[1] United Nations, *Basic problems of disarmament*, Reports of the Secretary General, New York, 1970.
[2] J.M. Tour, *Chem. Eng. News*, 78(28) (2000) 42.
[3] T.H. Nguyen, *Science*, 309 (2005) 1021.
[4] M. Mesilaakso, *Chemical Weapons Convention Chemicals Analysis*, John Wiley and Sons, Chichester, 2005.
[5] U.S. Congress, Office of technology assessment, *Technologies underlying weapons of mass destruction*, U.S. Government Printing Office, Washington, 1993.
[6] M.S. Meselson, *Sci. Amer.*, 222(5) (1970) 15.
[7] U.S. Dept. of Defense, Office of the undersecretary of defense for acquisition and technology, *The militarily critical technologies list*, Washington, 1998.
[8] M.R. Hurst, E. Wilkins, *Am. J. Appl. Sci.*, 2(4) (2005) 796.
[9] S. Hadlington, *Chem. World*, 3(9) (2006) 39.
[10] T.E. Mlsna, S. Cemalovic, M. Warburton, S.T. Hobson, D.A. Mlsna, S.A. Patel, *Sens. Actuators B: Chem.*, 116 (2006) 192.
[11] Y.C. Yang, J.A. Baker, J.R. Ward, *Chem. Rev.*, 92 (1992) 1729.
[12] J.F. Bunnett, *Pure Appl. Chem.*, 67(5) (1995) 841.
[13] C. Bismuth, S.W. Borron, F.J. Baud, P. Barriot, *Toxicology Lett.*, 149 (2004) 11.
[14] R.A. Greenfield, B.R. Brown, J.B. Hutchins, J.J. Iandolo, R. Jackson, L.N. Slater, M.S. Bronze, *Am. J. Med. Sci.*, 323(6) (2002) 326.
[15] World Health Organization, *Public health response to biological and chemical weapons: WHO guidance*, Geneva, 2004.
[16] G.W. Parshall, *Pure Appl. Chem.*, 74(12) (2002) 2259.
[17] M. Wheelis, *Pure Appl. Chem.*, 74(12) (2002) 2247.
[18] L. Szinicz, *Toxicology*, 214 (2005) 167.
[19] Congressional Research Service, *Small-Scale Terrorist Attacks Using Chemical and Biological Agents: An Assessment Framework and Preliminary Comparisons*, Government Printing Office, Washington, 2004.
[20] H. Nozaki, N. Aikawa, Y. Shinozawa, S. Hori, S. Fujishima, K. Takuma, M. Sagoh, *Lancet*, 345 (1995) 980.
[21] H. Nozaki, N. Aikawa, *Lancet*, 345 (1995) 1446.
[22] Stato Maggiore dell'Esercito, Ufficio NBC, n. 6210, *Aggressivi Chimici*, Rome, 1980.
[23] Swedish National Defense Research Establishment, *A FOA briefing book on chemical weapons*, Stockholm, 1992.
[24] NATO handbook on the medical aspects of NBC defensive operations, AmedP-6(B), 1996.
[25] Office of the Surgeon General, Department of the Army, U.S.A., *Medical aspects of chemical and biological warfare*, Washington, 1997.
[26] S. Kluge, L. Szinicz, *Toxicology*, 214 (2005) 268.
[27] F. Worek, M. Koller, H. Thiermann, L. Szinicz, *Toxicology*, 214 (2005) 182.
[28] A. Stone, *Science*, 297 (2002) 764.

Strengthening National Public Health Preparedness and Response to Chemical, Biological and Radiological Agent Threats. Edited by C.E. Cummings and E. Stikova.
IOS Press, 2007.

The Psychological Effects of Biological Agents as Terrorist Weapons

Massimo C. RANGHIERI [a,1], Matteo GUIDOTTI [a,b] and Alessandra ROSSODIVITA [c]
[a] *SMOM Auxiliary Corps of the Italian Army, 1° Reparto, Milan, Italy*
[b] *CNR-Institute of Molecular Sciences and Technologies, Milan, Italy*
[c] *Dep. of Cardiothoracic and Vascular Diseases, San Raffaele Hospital, Milan, Italy*

Abstract. Training programs for first responders do not adequately address the psychological effects of biological and chemical terrorism. Bioterrorism, being one of the most complex forms of terrorist attack, affects in remarkable ways the life of both survivors and first responders due to the very high impact of uncertainness of the consequences of the event. Cooperation with multidisciplinary experts such as psychologists, chemists, and toxicologists, will certainly enhance support and education for first responders.

Keywords. Biological warfare agents, NBC defense, terrorism, first responders, training program, psychological impact

Introduction

Training programs for first responders should consider:

1. The capacity for self-protection
2. The quick analysis of proper operating conditions
3. An understanding of the needs of the victims
4. The promptness of the entire operation

The psychological impact of terrorist events on first responders is too often neglected. This article highlights the importance of proper combined training programs where a remarkable emphasis is given to the monitor and control of panic attacks and related to psychopathologies.

Bioterrorism is the most complex and panic-inducing form of terrorist attack. The consequences of bioterrorism linger after the attack is actuated; the incubation time of viruses or bacteria allows the harm to continue after the actors disappear from the scene. Identification of the vectors may require time and the symptoms of attack may be initially perceived as normal illness symptoms.

1. The Needs of First Responders

An important and detrimental effect of terrorism is the emotional and psychological impact on first responders. While current training programs recognize first responders'

[1] Corresponding Author: EI – SMOM, 1° Reparto, via Saint Bon 7, 20147 Milano, Italy; e-mail: maranghi 2000@yahoo.it.

vulnerability to traumatic events, the focus is primarily on technical details of the effects of weapons of mass destruction and on how to help victims. Prevention methods should be developed to assist first responders' own mental health and coping strategies. An epidemiologic study [1] conducted by the University of Oklahoma found that 20% of the rescue personnel at the Oklahoma City bombing required mental health treatment. This demonstrates the need for qualified mental health professionals who can effectively identify and treat first responders.

In addition, there is also a need for further research on various treatment methods for first responders. For example, Critical Incident Stress Debriefing (CISD), a technique to assist field rescue personnel to cope with the stress of traumatic events, has gained widespread popularity [2,3]. CISD was originally developed as a relatively rapid technique to alleviate stress symptoms and prevent burnout among rescue workers. It involves organized group meetings for all personnel in the rescue unit (with or without symptoms), emphasizes peer support, and is led by a combination of unit members and mental health professionals. CISD has gained wide acceptance among field emergency workers and it is increasingly used with hospital-based emergency personnel, military service members, public safety personnel, volunteers, victims, witnesses, and even schoolmates of victims. It is strongly recommended that this technique be adopted in all cases where first responders have been providing assistance in bioterrorist related attacks [4].

2. Historical Cases

Chemical warfare agents (CWAs) have been used in several conflicts of the 20th century. Despite its widespread use, CWAs have generated more psychological casualties than physical injuries. For example, during World War I, there were more victims of fear than actual injury from gas. In fact, the ratio of psycho-pathology connected to gas fright and to physical injury was 2:1. This phenomenon is known as *gas hysteria*.

More recently, during the first Gulf War, Israel was the victim of several SCUD attacks that generated 1059 rescue requests. Of these, 233 requests were for direct victims and 826 were for indirect victims. Of the 826 requests for indirect victims, 544 were due to severe panic attack and 230 to the side effects of atropine (self-administration of atropine against supposed bombing by nerve agents). Overall, there were 11 fatalities, four from a heart attack and seven from the incorrect use of a gas mask. While these figures are low, they accurately portray the psychological consequences of an uncertain event [5].

Similarly, during the terrorist attack on the Tokyo underground, where Sarin was spread in the train and the station, there were 12 fatalities and 908 people were hospitalized. In total, first responders executed over 4023 rescue operations. More than 75% of these were related to psycho-pathology [5].

3. General Considerations

Terrorism is an action or a sum of actions aimed to reach a political, ideological or religious objective, through diffusion of panic, horror and fear of insecurity. Of all of the possible terror weapons, biological agents have a remarkable psychological effect. The simple announcement of their use generates horror and panic [6]. Behavior disorders not only affect the potential target victims, but also negatively affect both well trained and non-trained first responders.

The delay between bio-agent dispersion and its action prevents identification of contagion cause, expanding suspicion to all common daily activities. The use of effective tools and processes to avoid bio-agent diffusion, such as quarantine, is normally complex and disrupts the daily habits of individuals or of a population.

3.1. Reasons for Fear

Another reason for fear is the awareness that biological agents can be modified to enhance virulence or to reduce the therapeutic effect of known drug. For example, viruses or other pathogenic agent can be modified to become drug resistant.

By nature, some of biological agents cause symptoms identical or similar to trivial and common pathology. In turn, the delay in the proper identification of the "real reason" of the contagion allows a further distribution of the infection. Contagion may also occur through contact with pets and/or other animals living in surrounding farms.

The remote suspicion that an effective therapy does not exist will certainly increase a sense of impotence and generate anxiety. In all of these circumstances the role of the media is important. Media needs should never interfere with the proper flow of information. Further, information must be quick, precise and effective, never exacerbate alarm and panic.

In response to a bioterror attack, individuals first wonder what they should do. A sense of anarchy or a desire to not obey rules derives from the perspective that "all is lost, and there is nothing else to do". Helplessness and anxiety may ensue: the consequence of all of this is a perception of isolation, negation of reality, and fatalism.

4. Stressors

Figure 1 shows the various ways in which stressors may interfere with routine activities and cause psychosomatic symptomatology. Members of first responder teams should be trained to distinguish anxiety symptoms from organic pathology. For example, shortness of breath can be generated by lung infection or by chemical agents such as phosgene, but well-trained first responders will also understand that this symptom could indicate a panic attack. Multidisciplinary rescue teams supported by a mental health professional are strongly recommended to manage natural or man-made disasters [7].

The approach to treating victims of panic attack or an anxiety attacks, are quite complex and interest a remarkably broad spectrum of considerations. Sometimes only the specialist can obtain a satisfactory result. Risks factors for panic onset include the following:

- Perception that only limited amount of security resources is available,
- Clear perception of a very imminent risk,
- Perception of ineffectiveness of the available tools.

In the case of a bioterrorist attack, the perceptions are magnified in comparison to other terrorist scenario, increasing the risk of unestimated psychological consequences in the victims.

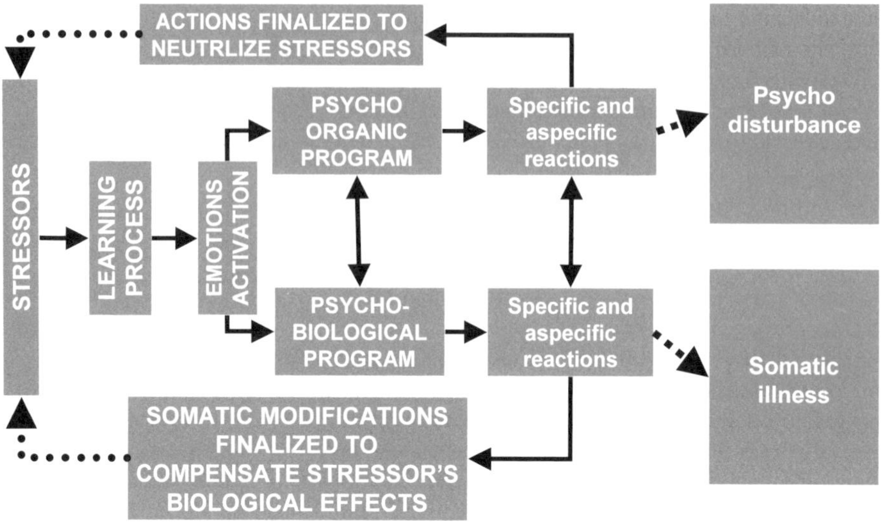

Figure 1.

Typical individuals exposed to the risk for psychopathology onset include:

- First responders' teams
- Paramedics and Physicians
- Team leaders
- Disaster managers

These groups very often need the support of a mental health professional.

Other groups that may require special attention among the "survivors" and where special care should be delivered include:

- Older adults
- Children
- Members of family where the event disaggregated the structure
- Media Operators

In all the above-mentioned cases, several psychological disturbances may occur, such as acute stress disorder (ASD), post-traumatic stress disorder (PTSD), drug abuse, anxiety, major depression, and psychotic disorders. Rescue teams must respond to minimize the effects of such disorders, and typical actions that will help address these issues include:

- The distribution of correct information to explain infection diffusion,
- The correct image of epidemiology,
- The risks connected to the infection,
- The description of a correct prophylaxis and a proper therapy.

With interdisciplinary resources, rescue teams should manage anxiety and panic and provide rapid triage and treatment. While this is true in general, in the cases of bioterrorism attack, it becomes of vital importance, particularly when the bio-agent mimics symptoms of common diseases.

Hyperactivation symptoms should always been monitored and controlled after the rescue action has been carried out, with realistic reassurance and soft sedation/medication if needed. As soon as the top of the effects slows down, the rescue team should re-build an effective social role within the victims and catalyze the reintegration process to re-address almost all previous daily duties the entire population interested to the event, were used to perform.

The organization of educational programs and meetings helps rebuild confidence and minimize symptoms associated with the event.

A rescue team leader should be prepared to effectively address the victims/population during a possible bio-attack, assuring them that fear and anxiety are normal human reactions to such an event. The team leader should know how to disseminate proper information from authoritative sources on the possible risks which may occur, remembering that the situation is often less critical then the media suggests.

For the self-psycho-defense of rescue teams, it will be important to find distraction to reduce tension, identifying activities that can be easily practiced. Members of rescue teams have to reduce the consumption of alcohol, caffeine or tobacco, and help victims in doing the same.

Conclusions

Compared with other weapons of terror, such as improvised explosives or chemical weapons, biological agents generate more apprehension and insecurity within the victims and the survivors. Special care should be taken by the first responders, who should always be helped by a mental health professional or psychiatrist. The proper management of information is vital to reduce stress and panic symptoms.

Current hospitals and medical structures response capabilities should be reviewed to update knowledge about chemical and biological warfare agents: this will allow the personnel to be better prepared to address the emotional symptoms that result from an attack with such agents.

While there is some knowledge about the psychological effects of bomb terrorism (the Oklahoma City disaster, for example) or of the consequences of natural disasters, little is known about the psychological effects that are specific to biological terrorism. This lack of knowledge further emphasizes the need for updating protocols and providing additional training of health providers and first responders to assure adequate mental health support in existing disaster networks.

Acknowledgements

M.C. Ranghieri and M. Guidotti wish to thank Prof. Rumm for the kind invitation to the NATO Advanced Study Institute held in Skopje (Macedonia), and the Italian Association of the Sovereign Military Order of Malta.

References

[1] S. Wessely, K.C. Hyams, R. Bartholomew, "Psychological implications of chemical and biological weapons", *BMJ* (2001), 323(7318), 878.

[2] G.B. Knudson, "Nuclear, biological, and chemical training in the U.S. Army Reserves: mitigating psychological consequences of weapons of mass destruction", *Military Med.* (2001), 166(12 Suppl), 63.

[3] J.A. Romano, "Psychological casualties resulting from chemical and biological weapons", *Military Med.* (2001), 166(12 Suppl), 21.

[4] H.C. Holloway, A.E. Norwood, C.S. Fullerton, C.C. Engel, R.J. Ursano, "The threat of biological weapons. Prophylaxis and mitigation of psychological and social consequences", *J. Am. Med. Assoc.*, (1997), 278(5), 425.

[5] T. Col. Med Braccini "Lectures at III Course for Army Medical Personnel" Army Medical School Cecchignola Dec. 2006.

[6] M.J. World, "Bioterrorism: the need to be prepared", *Clin. Med*, (2004), 4(2), 161.

[7] J. Jaax, "Administrative issues related to infectious disease research in the age of bioterrorism", *ILAR Journal* (2005), 46(1), 8.

8. Radiological Agents

Strengthening National Public Health Preparedness and Response to Chemical, Biological and Radiological Agent Threats. Edited by C.E. Cummings and E. Stikova.
IOS Press, 2007.

Medical Effects of Ionizing Radiation

Curtis E. CUMMINGS
Drexel University School of Public Health, MS 660, Philadelphia, PA 19102-1192, USA

Abstract. Questions about ionizing radiation eventually engage most health professionals. Because of their widespread use, radiation sources are an exposure risk and a terrorism threat. Acute radiation syndrome (ARS) occurs after exposure $\geq$ 1 Gray. Internal contamination occurs when persons ingest, inhale or are wounded by radioactive material. There is treatment for victims of radiation injury. ARS treatment includes supportive care, colony stimulating factors, and prophylactic and definitive treatment of infections. For internal contamination, decontaminate patients; give blocker, diluter and chelator drugs. Psychological effects will likely outnumber physical radiation injuries, and planners must account for their numbers.

Keywords. Ionizing radiation, acute radiation syndrome, internal radionuclide contamination

Introduction

Exposure events and concerns about radiation eventually engage most public health departments, and many clinicians and researchers. Because of people's fears, as well as real health threats, information about ionizing radiation is required for all health professionals.

1. Radiation Basics and Perspectives

Many stories are told about radiation, such as: "Coal-powered electrical plants are safer and emit less radiation than nuclear ones;" "Nuclear plants can blow up like atomic bombs;" "Any exposure to radiation is dangerous." All are false as stated, even the first! Coal plants release tons of radioisotopes in their smoke per year in the U.S., and they have many safety hazards. Radiation-related fables are common. Some believe that radiation creates monsters like Godzilla, or changes us in strange ways. Few know that we are exposed to radiation daily.

1.1. Basic Radiation Physics

Ionizing radiation is defined as: radiation energetic enough to create ions in atoms it strikes. It strips electrons from their orbitals, changing atomic structure. Types of ionizing radiation include alpha particles (poorly penetrating but an internal contamination hazard), beta particles (moderately penetrating; requires perhaps a sheet of aluminum for shielding); gamma and X-rays (can be highly penetrating; require lead or concrete shielding); and neutrons (can be fast and very energetic; produced by nuclear fission). Using the standard principles of time, distance and shielding, exposures can be minimized near powerful radiation sources.

1.2. Availability of Sources

Radio-isotopes are ubiquitous, including inside our bodies, and are widely used in industry and in medicine. There are tens of thousands of "large" gamma sources in the U.S., >10 Ci (curies) in activity, and hundreds of thousands of medium and small sources. Sources can be bought or stolen. The level of this threat is uncertain. The International Atomic Energy Agency recorded 18 cases of attempted nuclear trafficking from 2000 to 2005. In a May 2006 interview, Hans Blix, former U.N. Chief Weapons Inspector in Iraq, said that most such cases were "thieves and scam artists selling fake uranium and plutonium," that he did not worry about trade in weapons-grade material but that radiation dispersal devices (RDD's) were a threat. Accidental exposures are not rare. Settings where people have been exposed include:

- Abandoned medical equipment. Radiotherapy units have been sold as scrap metal. Their contents, containing large ^{60}Co and ^{137}Cs sources, spilled and caused serious radiation exposures and deaths, and area contamination.
- Industrial radiography. Strong gamma sources are used to radiograph metal hardware. These have been misplaced and caused exposures and deaths.

1.3. Cellular Chemistry and Radiation Effects

All forms of radiation injure cells the same way, differing only in how much damage each type causes. A particle or photon can directly hit a molecule, or damage can be indirect when radiation ionizes water, creating a shower of free radicals. Either way, molecular structure and function change instantly. Radiation breaks membranes, proteins, and most importantly, DNA. Enough DNA breaks cause cell death or long-term effects such as cancers. Malignancies associated with radiation include leukemias and solid tumors of the lung, thyroid, breast, liver, bone, intestines, oropharynx, nasopharynx, skin, and other tissues.

Radiation's effects depend on the amount of energy deposited/ absorbed by the body – the dose – and the distribution of the deposited energy – whole-body, partial body, external, or specific organ. There are 3 main routes of exposure, or modes of injury, for ionizing radiation:

1. External penetrating radiation exposure, causing acute radiation syndrome (ARS).
2. Local injury, when part of the body is externally exposed, e.g., by a collimated beam.
3. Internal contamination with radionuclides. (All can occur at once.)

1.4. Dose Units

- Roentgen (R) – X/γ-ray exposure (not a dose unit; older term); measures ionization of air.

- Rad (R – "radiation absorbed dose") – traditional unit of absorbed energy to ionize any material, including human tissue. (Lab formula: 1 rad = 100 ergs absorbed/gm.) Low dose.
- Rem (R – "roentgen equivalent man") – effective dose [(rad) x (weighting factors: tissue, radiation type)]. (For photons, a rem, rad and roentgen are equivalent; all are called "R".)

In current International Units (SI units):

- Gray (Gy) – SI unit of absorbed dose. 1 Gy = 100 rad = 100 cGy (= 1 J/Kg)
- Sievert (Sv) – SI unit effective dose = 100 Rem

Other terms exist to measure radioactivity, such as the curie and becquerel.

The average whole-body dose to the average American is 360 millirem (mR) per year. About 82% is natural radiation, including natural radioisotopes in the body. Over half of natural radiation, 200 mR, is from radon that accumulates in buildings. Figures vary by geography (different content of soil and bedrock) and altitude. Of the man-made fraction, 3% is from consumer products, including cigarettes. Legal standards for radiation exposure in the U.S. are 0.001 Sv/ year for the general public and 0.05 Sv/ year for radiation workers [1]. This level protects workers, and no ill effects are found in their bodies on medical testing.

2. Acute Radiation Syndrome

ARS is an acute illness that occurs as a combination of clinical signs and symptoms, in stages, over hours to weeks after "prompt" or high dose rate exposure to penetrating radiation in a dose of $\geq$ 1 Gy (100 R). It requires irradiation of the whole body or most of it and it evolves predictably as injury to tissues and organs is expressed. External exposure is the usual cause, although there have been rare cases of ARS after internal contamination with gamma sources.

2.1. Pathophysiology

Radiation causes two kinds of microscopic pathology that lead to clinical features of ARS – depletion of stem cell lines, and microvascular injury. At survivable doses, stem cells lines can regenerate. Primitive and rapidly dividing cells are the most radiosensitive cells, are injured the most by radiation and chemotherapy, and are (in decreasing order): spermatogonia, lymphocytes and oocytes, erythroblasts, other hematopoietic tissue, intestinal crypt cells and hair follicles (sterility, anemia, diarrhea and hair loss involve these tissues). Vascular injury causes the clinical effects of whole-body doses > 5 Gy and of local and skin injury.

At doses of 0.1 to 0.7 Gy clinically detectable effects are asymptomatic; the annual legal occupational exposure limit, U.S., 0.05 Gy, is less than ½ of the level of any detectable effects. At 0.12 Gy, sperm count decreases; at 0.20 Gy there is an increase in abnormal chromosomes; from 0.20 to 0.70 Gy, bone marrow depression with lymphopenia is detectable. For whole-body irradiation the LD_{50} with no medical treatment is about 3.5 Gy. The LD_{50} is 5 Gy or more with current medical treatment [2].

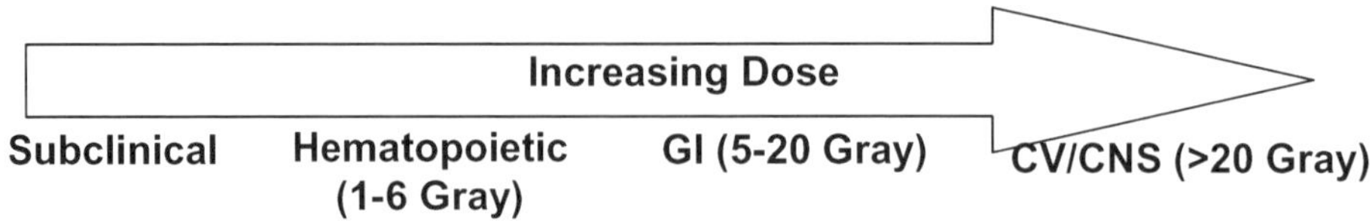

Figure 1. Increasing radiation effect with increasing dose – ARS sub-syndromes.

ARS is actually a series of illnesses that each evolve over time, with increasing dose – the hematopoietic, gastrointestinal (GI), cardiovascular (CV), and central nervous system (CNS) subsyndromes (Fig. 1). They are progressive and additive with rising dose. The CV and CNS are combined since they are rapidly lethal and vascular injury causes both.

2.2. Clinical Stages

A case of ARS displays a series of time-dependent stages. Higher doses shorten the time to onset and increase signs and symptoms. There is individual variation in timing, severity, and if the patient has other disease or injury. In the prodrome, chief symptoms are nausea and vomiting, with malaise, fatigue and weakness. Release of cell mediator chemicals such as histamine, interleukins and cytokines, and free-radical effects act at the brain's vomiting center and cause this non-specific clinical picture. In the latent period, most symptoms subside; fatigue and weakness may remain. Because of the immune suppression, any major injury becomes a combined injury, with high mortality, so patients must be ordered to "no field duty" and assigned benign tasks such as office work. Manifest illness stage – clinical signs and symptoms are associated with the major organ system injured (blood-forming elements, intestine, brain). Recovery stage or death – survival decreases with increasing dose, and sepsis is the usual lethal event, due to hematopoietic failure [2–4].

2.3. Hematopoietic Syndrome

The classic dose range stated in textbooks is "1 to 5 Gy," but it begins at about 0.7 Gy and continues at higher doses. Stem cell lines in bone marrow are injured, so all blood cell line production is reduced or stopped. Pancytopenia will occur above about 2 Gy.

Clinical stages – Prodrome onset is 3 to 16 hours; duration is $\leq$ 48 hours. Symptoms are nausea, vomiting and anorexia in most patients; diarrhea may occur towards 5 Gy; symptoms are treatable. The 3 to 4 week latent period is asymptomatic except for possible mild fatigue, anorexia and hair loss. Onset of manifest illness is at 3 to 5 weeks, with severe pancytopenia. It is survivable if the bone marrow is resuscitated and infections and hemorrhage are prevented. Infections may show few signs and symptoms because of the immune suppression.

The most useful laboratory test is the complete blood count (CBC) with absolute white blood cell (WBC) counts. A series of CBC's, if graphed (Fig. 2, dose of 3 Gy), shows typical changes. Lymphocyte counts fall starting on day 1; other WBC's and platelets follow. Erythrocytes may remain in or near the normal range. At 4 weeks there is a life-threatening clinical picture if untreated, as cell lines bottom out. Without WBC's, overwhelming infection is likely, and without platelets there will be hemorrhage [2–4].

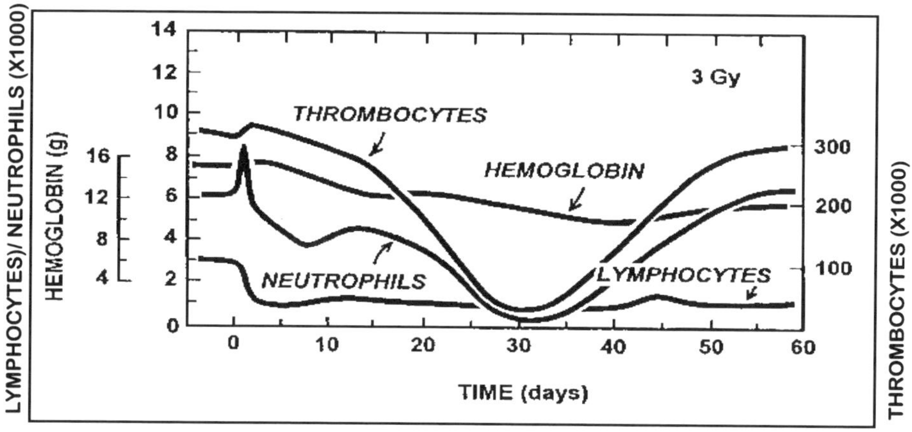

Figure 2. Hematological response to 3-Gy whole-body exposure, over a 60-day period.

2.4. Gastrointestinal (GI) Syndrome

The classic range is 5 to 20 Gy; the full-blown GI syndrome above 8 Gy has always been fatal [4]. Hematopoietic syndrome will occur concurrently, if patients survive long enough. The GI tract injury involves both dividing cells and small blood vessels. With 5+ Gy dose, crypt cells die within 2 days, and with no re-supply of cells, villi shorten and slough off. Small vessels break down and swell, leading to blood stasis and bleeding [4].

Clinically, a severe prodrome begins usually in 1 to 2 hours, lasting 48 + hours, with 100% of victims having intractable nausea and vomiting, diarrhea and fever. In the 5 to 7 day latent period, malaise and weakness are severe enough to be disabling. There is no special blood test or other clinical sign to track the GI syndrome, but CBC's are needed anyway, because of the bone marrow injury. The manifest illness is stormy, with several severe events at once, any of which can be the terminal event. Mucosal break-down can cause a paralytic ileus, with abdominal distension, vomiting, diarrhea and GI collapse, and fluid and electrolyte shifts with cardiovascular compromise, then shock. Damaged vessels can hemorrhage. Bacteria translocate across damaged mucosa leading to sepsis, the usual cause of death.

2.5. Cardiovascular (CV)/Central Nervous System (CNS) Syndrome

Beginning at 20 Gy, the full syndrome occurs above 50 Gy. Such extreme doses have resulted from nuclear fuel handling accidents, where a critical mass suddenly formed, termed a "criticality event". Pathologically there is microvascular injury with increased endothelial permeability, especially in the brain, and resultant microvascular infarcts and cerebral edema.

The prodrome begins within minutes to 1 hour, with severe nausea, explosive vomiting and diarrhea, and CNS signs such as epileptic seizures and altered mental status. Erythema of the skin (a blush, not a burn) is from the endothelial injury and cell mediators acutely released. The latent period is brief, several hours to 1 to 2 days. Victims may be lucid and in no pain, although weak. The manifest illness phase is rapid and inevitable, with deteriorating CNS status, reduced consciousness, seizures, diarrhea, respiratory distress, and uncontrollable swings in systemic blood pressure. Coma and death from cerebral edema occur in 2 to 3 days.

2.6. Cutaneous Radiation Syndrome (CRS)

Radiation skin injury has been caused by radiotherapy, by skin contamination by beta emitters, and after direct contact with energetic sources. It occurs alone or with whole-body exposure. Damage to stem cells in skin and hair follicles and to small vessels causes CRS. Thermal and chemical burns may look similar, but their time course (immediate) and pathology are very different. It acts as a sub-syndrome, with the same clinical stages as other sub-syndromes.

The prodrome is a wave of erythema within hours of exposure, peaking at 24 hours, lasting up to 48 hours. Not a burn, it is a skin blush as previously mentioned. The latent period is clinically quiet. Manifest illness, onset at 10 to 14 days, features hair loss, intense erythema, scaling and desquamation. Subepidermal blisters, ulcers, and moist desquamation can develop; these are unpleasant, painful and indicate higher doses. Recovery, chronic radiation dermatitis, begins at about 2 months. Features include radiation fibrosis (very painful, leading to vascular occlusion with secondary ulceration) and telangiectasias, a characteristic sign.

2.7. Dose Assessment (Diagnosis)

Two crucial clinical steps are triage and dose assessment. Perform triage as in all medical emergencies, treat life-threatening injuries, then do any required decontamination (likely needed in a nuclear setting).

Initial diagnostic steps. ARS patients do not arrive for treatment wearing tags with dose numbers on them! Dose must be calculated and revised over time, as clinicians get data. The best two, and fastest two steps are: time to onset of vomiting, and lymphocyte depletion.

- Prodrome. The timing of the prodrome, especially the time to onset of vomiting, suggests a dose range. Observe the onset, duration, and subsequent latency, record these as data, then decide what dose they mean and plan treatment [2,4].
- Lymphocyte counts. These are the best rapid gauge of dose, dropping quickly with high doses. Get CBC's with absolute WBC counts, every 4 hours on Day 1 and daily thereafter. The classic Andrews figure [5] is still in use. A 50% drop in 24 hours means a serious radiation injury, Curve 2 or 3, and a further drop indicate a lethal dose.

Later data – physical dosimetry, biodosimetry, appearance of CRS. These data are gathered over time. For individuals, personal dosimeters are useful, but patients may not be wearing them or they can get damaged. Dosimetry may be available from combat units that have health physicists and equipment. Biodosimetry is performed at Armed Forces Radiobiology Research Institute (AFRRI), at Oak Ridge National Laboratory and at some universities. On a standard karyotype (chromosome smear), a re-

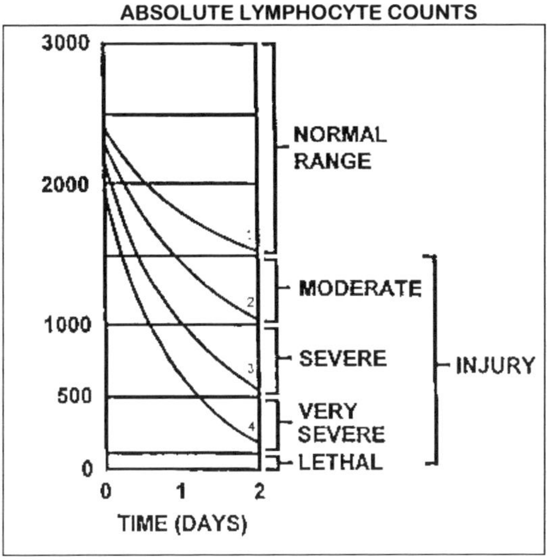

Figure 3. Lymphocyte depletion curves (Andrews, 1965); relationship between early changes in peripheral blood lymphocyte counts and degree of radiation injury. Curve 1 – 3.1 Gy; curve 2 – 4.4 Gy; curve 3 – 5.6 Gy; curve 4 – 7.1 Gy.

searcher counts the chromosome abnormalities, especially those with 2 or more centromeres (dicentrics). This yields a good dose estimate in 3 days. Newer methods can be done in 3 hours using PCR and immunofluorescence techniques. All use WBC's from a blood sample, sent in a standard medical laboratory blood tube anti-coagulated with heparin, preferably lithium-heparin [4]. CRS, as previously stated, requires a minimum skin dose of 3 Gy, and the more CRS signs that develop, the higher the dose was. Assessing dose is a process, over time, that should include as much data as possible. Health physicists have said, "Eventually, the patient will tell you the dose."

2.8. Treatment of ARS

Hospitalize ARS patients at the onset of the manifest illness phase or when CBC's indicate significant immune suppression. Hospitalization may not be necessary in the latent phase. With treatment, most uncomplicated cases survive a dose of 5 Gy or less.

Supportive care. First, maintain vascular integrity. Stop fluid losses, including by use of anti-emetics. Above about 1.5 Gy phenothiazines are ineffective, so use 5-HT3 inhibitors such as ondansetron (Kytril™) or granisetron (Zofran™). Control of vomiting can reduce the risk of a patient being injured because he was temporarily debilitated. As needed, administer IV fluids and blood products including platelets. Prevent hemorrhage by transfusing if platelet counts are < 20,000; treat it with any count if there bone marrow injury and bleeding. When WBC counts drop, put patients into standard "clean rooms", as for oncology or HIV patients, with reverse isolation, gowns/gloves/masks – the usual methods.

Antibiotic prophylaxis. Most important is gut decontamination against gram negative bacilli, begin oral fluoroquinolones such as ciprofloxacin within 1 to 2 days of exposure. For gram-positive coverage, add amoxicillin or a macrolide. As in oncology and HIV patients, give antiviral prophylaxis when the absolute neutrophil count (ANC) falls. Against cytomegalovirus (CMV) use ganciclovir or acyclovir, in CMV positive patients. Use CMV-free blood products, in CMV negative patients. Against herpes, use acyclovir, in patients who meet Walter Reed criteria (titers > 1:8, or frequent recurrences). For antifungal treatment, fluconazole gives superior results in oncology patients.

Definitive treatments. Specific treatments include: antibiotics to treat infections that do occur, and colony stimulating factors (CSF's – synthetic cytokines) for bone marrow recovery.

In general, care of infections in ARS patients is similar to care of oncology patients with neutropenic fever. Treat if there are signs of infection (e.g., fever) with ANC < $500/mm^3$. An infection focus may not be obvious, so culture all usual sites of infection. Then, begin broad-spectrum IV antibiotics. Two options recommended by AFRRI microbiologists and U.S. Navy infectious disease physicians are:

- Ciprofloxacin/levofloxacin $\pm$ amoxicillin/vancomycin
- $3^{rd}/4^{th}$-generation cephalosporin $\pm$ amoxicillin/vancomycin

Antibiotics in ARS patients must preserve the normal GI tract flora, unlike in oncology patients, so do not give antibiotics that kill anaerobic bacteria. When a specific pathogen is cultured, treat it. If fever persists for 5–7 days despite treatment, add antifungal coverage. Continue treatment until the ANC is >500, even if the fever breaks.

CSF's have saved lives of ARS patients by stimulating WBC recovery. The most widely used is Neupogen™, G-CSF (granulocyte colony-stimulating factor), which is now in the U.S. Strategic National Stockpile (SNS) (dose: 2.5 to 5 mcg/kg daily s.c.). A long-acting G-CSF is Neulasta™, PEG-G-CSF (dose: 300 mcg/kg s.c. once). Both are U.S. FDA-approved for use in cancer patients. Start these as soon as ARS is diagnosed, within 24 hours if possible. Continue until there is a neutrophil spike that indicates bone marrow recovery [2,4].

3. Internal Radionuclide Contamination

3.1. General Information

Health physicists define contamination as "unwanted radioactive material present on or in the body" [4]. Internal contamination occurs when unprotected persons inhale or ingest radioactive material, or material enters the body through wounds, or is administered medically. Contamination can be external, internal, or both. Most contaminants are solid particulate matter that is on the ground, or are aerosols.

Acute medical effects of contamination are none; patients are initially asymptomatic. The main medical concerns are chronic effects – injury of target organs (e.g., lung, bone), and malignancy. Rarely, internal contamination with high-energy isotopes has led to ARS.

With the wide use of radionuclides, episodes of external and internal radioisotope contamination episodes have occurred regularly since tracking began in the 1940's. Medical errors during diagnosis and treatment have been the most common cause of significant internal contamination, with accidents second [3,6]. The most common such accidents have been industrial accidents. Commercial high-energy sources (e.g., ^{60}Co, ^{137}Cs and ^{192}Ir) used industrial radiography, commercial irradiators or medicine can readily cause injury. Although dangerous, these sources are now usually encapsulated and less likely to cause contamination. "First-responders" can easily locate a source using RADIACs, and can then avoid the hazard, don protective gear and make a safe perimeter. To make an RDD, terrorists can use any particulate radioisotope, preferring energetic isotopes in powder form that can be disbursed [2,3]. In such an event, a health physicist or an NBC team can take samples and identify the isotope using GC-mass spectroscopy or other specialized equipment.

3.2. Pathophysiology and Chemistry

Radionuclides obey the same principles of toxicology and pharmacology that non-radioactive (stable) toxins and chemicals obey. Factors that determine the amount of internal hazard are the amount of the radionuclide(s); the type of radiation emitted and its energy; the length of time in the body (biological half-time); and the critical organ affected.

Biological half-time is: the time in which half the atoms of a substance are removed from the body. This number includes both the physical half-life and metabolic clearance. With this number and the amount of nuclide, clinicians can plan treatment, using standard references [7]. A radioisotope's life cycle in the body involves intake, uptake, deposition and elimination. Routes of intake are (in order of efficiency) inhalation, ingestion and skin penetration.

A radioisotope is chemically identical to a stable isotope of the same element. Both are metabolized according to their chemistry, which are the same, so an isotope's critical organ is determined by chemical and physical properties. Radionuclides distribute whole-body if the stable version is found whole-body. Examples are ^{24}Na, ^{137}Cs (which mimics K+) and tritium (which acts as water). "Bone-seekers" mimic calcium; such as radioactive lead, Sr and Ra.

3.3. Initial Patient Evaluation and Management

As with ARS patients, emergency care is the same as with cases unrelated to radiation injury – since contamination is not a medical emergency and causes no acute medical effects. Next, evaluate for contamination of skin and wounds. If there is external contamination, suspect internal contamination. Start treatment for internal contamination when the patient is medically stable and after external decontamination. Hospital staff and first responders may fear such patients, but proper technique and personal protective equipment prevents exposure. U.S. medical personnel have never had a major exposure while treating radiation victims.

Initial evaluation should begin at the accident scene if possible, or just after emergency care is complete. It includes the history of the accident, initial area and patient surveys, and nasal swabs. As in all of medicine, diagnosis of contamination is largely by history; the patient or another worker may know what isotope was being used. Area and patient surveys use a standard RADIAC with both beta-gamma and alpha probes. Collect a smear sample, or "swipe" any site and any wound that has a high count, and swab the nares [7].

External decontamination step reduces intake, and is similar for all NBC agents. It is simple in concept: Remove all clothing. Wash the patient (or have them take a shower). These steps remove 95%+ of external contaminants immediately. Use any mild detergent, assess for residual contamination, repeat if needed [3]. For persistent external contamination, abrasive soaps (used carefully) or heavy-metal chelating agents such as EDTA may be used.

Measuring internal contamination – methods include direct measurement with machines, and indirect measurement of samples of body fluids and excreta. A whole-body counter (most reliable; limited availability) or hand-held RADIACs directly measure the body's radioactivity. Swabs and swipes should already be done (as above). Bioassay is indicated to detect and quantify some isotopes in urine or feces.Obtain a series of spot urine and/or 24-hour urine and stool collections samples post-exposure, according to standard references [7].

3.4. Medical Treatment

Begin treatment as soon contamination is confirmed or the history is compelling. Blocking agents saturate a critical organ with a stable isotope, reducing deposition of the nuclide. For ^{131}I, use KI; for adults: 130 mg daily (duration of dose depends on duration of exposure) [8]. Diluting agents are used to flood the body with a stable isotope, reducing nuclide uptake in a critical organ. Tritium is diluted by stable water; treatment is to force fluids orally [7]. Chelating agents are ion-exchange drugs that mobilize toxic metal ions from the circulation. The kidney excretes the stable complex, chelator and toxin together. Chelation is effective for classic poisoning by some stable heavy metals, and for their radioisotopes. The DTPA salts (diethylenetriamine pentaacetate trisodium calcium/ zinc) are the most effective drugs to remove transuranic isotopes (i.e., plutonium, americium and curium). Approved by the U.S. FDA, they are in the SNS, and no serious toxicity has been reported. Dosage is 1 gram IV daily [3]. Prussian Blue, the common pigment, is an effective oral treatment for radioactive Cs, Tl and Rb, in a dose of 1 to 3 grams, three times daily. Lists of other chelators, blockers and diluters are in standard references [7].

4. Psychological Effects

The psychological effects associated with radiation are more important than ever, and will likely give rise to the greatest numbers of casualties. People have spoken, and the media have written, of thousands of deaths from Chernobyl soon after the accident, but actual figures from the first 3 months were: 31 deaths and 300 hospitalized due to acute injuries and ARS. Psychological effects, however, were indeed widespread and severe. Despite low exposures, typically below 1 R, many in Europe decided that they or their children would be injured by the radiation. As a result, the WHO has estimated that there were 150,000 to 175,000 excess abortions, and 5000 to 10,000 excess suicides, in the 6 months after the accident [9]. These may be considered psychological casualties. By these figures, the ratio of psychological to physical casualties was about 500 to 1.

(There are no good figures for the many thousands of cases of depression and post-traumatic stress disorder (PTSD) potentially making the ratio higher.) Similar patterns emerged after other publicized accidents [10].

History indicates that certain psychological injuries occur after radiation accidents: acute stress disorder (ASD), PTSD, suicides, and chronic fatigue syndrome. "Panic" has not been described after radiation accidents, and is rare. Many others will have single symptoms such as severe insomnia and anxiety that fall short of an ASD diagnosis yet are disabling [10].

An RDD attack can have major psychological effects. In a covert attack upon a large public venue such as a theme park or stadium, a radioisotope could be spread among maybe 100,000 people who walked through material placed on the ground. Doses would likely be small, 1 to 50 mSv. The usual rate of death from cancers is about 25%, the low-dose exposure might cause a calculated 1 to 5 additional deaths, yet many would blame their cancers on the attack. The psychological impact will be huge. People will not feel well, with symptoms they will attribute to "radiation." The economic effects would similarly be great.

For treatment, the military model is effective, including among civilians. Psychological effects of a disaster can be treated using the principles "P.I.E." – Proximity, Immediacy and Expectancy. This means: treat stress injuries nearby (proximity), do it quickly, and expect them to return to duty quickly, and tell them so. This prevented over 60% of disability-related behavior in Israel after the 1991 Scud attacks, and with the Yom-Kippur War [11].

5. Summary Points

1. Contrary to popular belief, there is treatment for victims of radiation injury. We can and must treat them. So, suspect that there is a radiation risk, and be ready to diagnose and treat.
2. For all radiation-related injuries, treat conventional injuries first – cases of ARS and internal contamination are not immediately life-threatening.
3. Definitive treatment of ARS includes proper antibiotic choice and CSF's. For internal contamination, perform decontamination; assess internal contamination by direct and indirect measurement; administer blockers and chelators early.
4. Psychological effects are likely to outnumber other radiation injuries, and the health care system must have the surge capacity for such patients.
5. Patients are manageable and can be helped, if there is a good plan and a system ready to handle them. Public health and health care systems must be ready for radiation events. Medical planners must address staffing and supply requirements for the patients.

References

[1] Title 10, Code of Federal Regulations, Part 20, USA.
[2] Mettler FA & Voelz G. (2002). Major Radiation Exposure – What to Expect and How to Respond. *New Engl Jour Med* 346 (20), 1554–1560.

[3] *Medical Management of Radiological Casualties Handbook*, Second Edition, Armed Forces Radiobiology Research Institute (AFRRI), Bethesda, MD, 2003.
[4] Gusev IA, Guskova AK, & Mettler FA, editors. *Medical Management of Radiation Accidents*, Second Edition, CRC Press, Boca Raton, 2001.
[5] Andrews GA, Auxier JA., & Lushbaugh CC. (1965). *Importance of dosimetry to the medical management of persons accidentally exposed to ionizing radiation.* In: *Personnel dosimetry for radiation accidents*. International Atomic Energy Agency (IAEA), Vienna.
[6] U.S. Radiation Accident Registry. Radiation Emergency Assistance Center/Training Site, Oak Ridge Institute for Science and Education, Oak Ridge, Tennessee.
[7] National Council on Radiation Protection and Measurements (NCRP) Report No. 65, *Management of Persons Accidentally Contaminated With Radionuclides*. Bethesda, MD, 1979.
[8] U.S. Food and Drug Administration. *Guidance Document: Potassium Iodide as a Thyroid Blocking Agent in Radiation Emergencie*s. Available at: http: www.fda.gov/cder/guidance/4825fnl.htm.
[9] Bard D, Verger P, & Hubert P. (1997). Chernobyl, 10 years after: health consequences. *Epidemiol Rev* 19: 187.
[10] Collins D. (2002). Human responses to the threat of or exposure to ionizing radiation at Three Mile Island, Pennsylvania, and Goiania, Brazil. *Mil Med* 167 Suppl 1: 137.
[11] Pastel R. (2002). Radiophobia: Long-term psychological consequences of Chernobyl. *Mil Med* 167 Suppl 1: 134.

9. Special Topics

Strengthening National Public Health Preparedness and Response to Chemical, Biological and Radiological Agent Threats. Edited by C.E. Cummings and E. Stikova.
IOS Press, 2007.

A Framework to Understand the Centrality of Protection and Restoration of Ecosystem Services to Water Management and Preparedness: An All-Hazards Approach with Implications for NATO Plans and Operations

Conrad Daniel VOLZ, DrPH, MPH [a,b,c,1]

[a] *Scientific Director, Center for Healthy Environments and Communities (CHEC)*
[b] *Assistant Professor, Department of Environmental and Occupational Health, Graduate School of Public Health (GSPH), University of Pittsburgh*
[c] *Co-Director for Exposure Assessment and Control; University of Pittsburgh Cancer Institute, Center for Environmental Oncology (UPCI-CEO)*

Abstract. Degradation of ecological services including deforestation, wetland loss, loss of topsoil and plant cover, loss of riparian habitat and natural drainage patterns, and stream and land erosion can have either a direct or indirect effect on both the quality and quantity of water resources. There exists a diverse literature on the contribution of a healthy, functioning ecosystem to a sustainable hydrological cycle, adequate aquifer recharge, water purification, biodegradation of human and animal toxins and prevention of storm surges and flooding. However a search yielded no publications relating to the use of ecosystem services as a primary public health prevention strategy to prepare for acute or chronic water management emergencies. Water scarcity or the contamination of water both intentionally or unintentionally can cause immediate and/or intermediate and/or long-term social and political upheaval and should be of great interest to NATO in its quest for peace.

This manuscript characterizes the interplay between water and land management issues leading to problems of water quality, quantity, stormwater surges and flooding that are a direct threat to public health. A framework to understand the complex chain of causation leading to tertiary public health outcomes is presented. The importance of protection and restoration of ecological services, as a primary public health prevention strategy to break the chain of causation will be described. Finally, an all hazards approach to water management will be incorporated and stressed throughout the narrative. It is argued that the extent to which tertiary water problems, such as flooding deaths and property damage or increased risk of human development of water related diseases, are expressed can be mediated by the vibrancy of ecosystem function.

Keywords. Water Management, Ecosystem Services, Public Health Preparedness, Peace and Conflict Studies, Water Quality and Water Quantity

[1] Corresponding Address: University of Pittsburgh, Department of Environmental and Occupational Health (EOH), Graduate School of Public Health, A712 Crabtree Hall (PUBHT), Pittsburgh, PA 15261; e-mail: cdv5@pitt.edu.

Problem Background: The Need for Water Management Preparedness

Water management is the most important global public health dilemma of the 21st Century. Urbanization, population expansion, industrialization, source contamination, watershed and habitat destruction and agriculture have placed severe strains on both surface and groundwater sources. In arid and semi-arid areas of the developed and developing world, water is being removed from fossil aquifers at unsustainable rates. Overuse of impounded and diverted surface water for agriculture is responsible for soil salination and decreasing crop yields. Groundwater pumping is also responsible for loss of riparian and wetland habitats, intrusion of saltwater and the movement of toxic and carcinogenic substances from contaminated vadose zones into potable water supplies. Worldwide, pathogens in water remain a central public health issue they are widespread, endemic and epidemic. In the USA, waterborne pathogens have caused epidemics largely via wet weather events causing sewer overflows and runoff combined with municipal drinking water treatment failures.

The USA is facing a national water crisis, which has been termed "the freshwater imperative" (Naiman et al., 1995); water quantity and quality issues are important internal as well as external national security threats. Throughout the world water resources are the basis for many inter and intra-governmental armed conflicts and have been the focus of regional political problems within many countries. NATO countries have a strategic interest in water management in areas of the world where political instability is related to water quality and quantity.

A Definition of Water Management

Water management is a term used to describe, holistically the intersection of water issues including; water sources and treatment; sewage and wastewater treatment; water contamination by toxic and carcinogenic chemicals, metals and radionuclides; water quantity; agricultural uses of water; stormwater and drainage; flooding; and watershed protection and associated development and transportation project considerations. This is a non-traditional public health model because customarily these issues are treated separately and the problems associated with agricultural appropriation of water or watershed protection and restoration are relegated to the sciences of land use planning, agricultural science and/or ecology.

While each facet of water management goes together to form an interlocking whole, single issues should be explored in depth in order to focus more attention on solutions. Caution needs to be exerted because the effects of solutions for a specific problem need to be reexamined using the holistic water management model. Sometimes solutions to one problem can exacerbate another problem, as installation of stormwater drainage systems (moving vast quantities of water away from developed areas) make downstream flooding at stream pinch point far worse. The development of long term, integrated watershed management plans that deal with the primary environmental, social and economic causes of water problems are necessary for economic stability and environmental sustainable. Environmentally sustainable communities and regions are the building blocks for national unity and stability.

Water management concerns vary according to geography and climate. Thus water management in relatively rainfall and surface water rich Central Europe will differ from the concerns of the Middle East where extensive groundwater pumping and use of

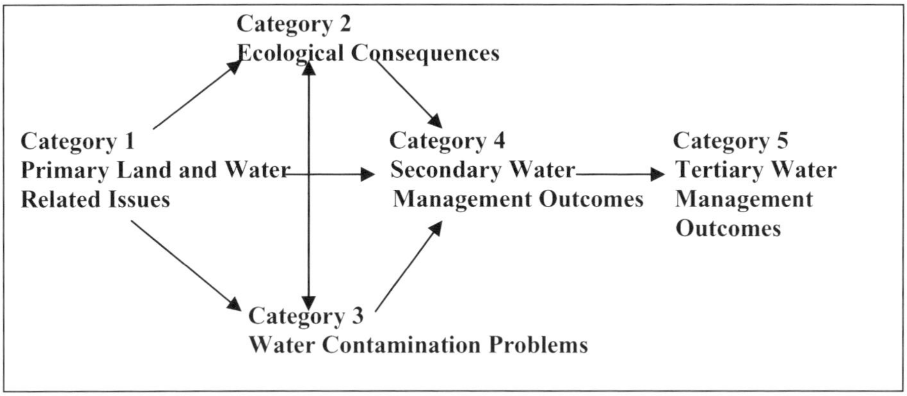

Figure 1. A Chain of Causation: Primary Land and Water-Related Issues to Tertiary Public Health, Social, Emotional and Economic Outcomes.

impoundments are employed for drinking water and agricultural usage. However the primary and secondary issues that ultimately cause environmental public health problems are very constant throughout the world and only vary in relative importance. For instance increased agricultural use of marginal slopelands in Highland Chiapus, Mexico leads to degradation of ecosystem services, soil erosion, high sediments in water and downstream flooding. This is not unlike development in critical watersheds in Central Europe, which have been implicated in exacerbation of flooding as highlighted by major loss of life and property damage in 2002.

How Do Water Related Primary Environmental, Social and Economic Issues Lead to Tertiary Public Health, Social, Emotional and Economic Outcomes

Figure 1, A Chain of Causation: Primary Land and Water-Related Issues to Tertiary Public Health, Social, Emotional and Economic Outcomes presents a basic framework to understand how environmental, social and economic issues related to water/land management (since land management issues have a direct and sometimes immediate effect on water outcomes and vise versa the term water/land management will often be used in this manuscript) can lead to significant environmental public health, medical, social, behavioral-emotional and economic problems. Headings for the major categories correspond to the public health paradigm of primary, secondary and tertiary prevention. This purposeful analogy was included to show where interventions can most successfully be applied to break the chain of causation. Briefly, primary land and water related issues alone or in combination with degraded ecological services and/or existing contamination problems cause secondary water management outcomes. These secondary outcomes such as increased human exposure to pathogens or toxins lead directly to tertiary environmental public health problems of increased risk of development of disease and actual disease states. It is more cost effective and less injurious to the public to prevent public health problems at the primary level. Depicting the complexity of interactions of all possible social, economic, geological, hydrological, biological, ecological, fate and transport and environmental dimensions is beyond the scope of this work. Readers who want more detailed information can consult the National Research

Table 1. Category 1: Primary Land and Water –Related Social, Environmental and Economic Issues

Legacy and Ongoing Industrial Pollution
Intentional Releases of Toxic/Hazardous/ Substances
Spills/Accidental Releases of Toxic/Hazardous Substances
Abandoned/Active Mines
Open Sewers and Failing On-Lot Septic Systems
Aging/Inadequate Municipal Sewer Infrastructure
Overuse of Impounded and Diverted Water for Agriculture
Agriculture in Marginal or Ecologically Sensitive Land
Past and Present Agricultural Fertilizer/Herbicide/Pesticide Use
Lack of Coordinated Water/Land Management Plans
Inappropriate Municipal Waste Disposal
Development in Headwaters and Critical Watersheds
Inappropriate Transportations Projects
Power Plant and Industrial Air Emissions/Deposition and Transport in Water
Lack of Basic Water Sanitation
Lack of Municipal Water Interconnectivity
Climate Change
Treating Surface and Groundwater as Not Interconnected
Attitudes/Behaviors Concerning Unlimited Water Use

Council (NRC), Committee on Watershed Management's New Strategies for America's Watersheds (NRC, 1999) and Sources, Pathways and Relative Risks of Contaminants in Surface Water and Groundwater: A Perspective Prepared for the Walkerton Inquiry (Ritter et al., 2002).

Primary Water Related Environmental, Social, and/or Economic Issues and "Downstream" Ecological and Water Contamination Consequences

Category 1 water-related environmental, social, and/or economic issues are quite varied and the importance of each factor for different regions of the world again depends again on climate and geography but also infrastructure development, political organization, industrialization and. technological advancement. Category 1 issues are detailed in Table 1, Category 1: Primary Land and Water-Related Social, Environmental and Economic Issues. Category 1 issues can create or exacerbate both category 2 ecological consequences/issues (Table 2, Category 2: Ecological Consequences) and category 3 water contamination problems (Table 3, Category 3: Water Contamination Problems). So category 1, 2 and 3 problems act as a triad to produce further public health problems. The category 2 consequences of loss of ecological services including wetland loss, deforestation, loss of topsoil and plant cover, loss of natural drainage patterns, stream and

Table 2. Category 2: Ecological consequences

Wetland Loss
Deforestation
Loss of Topsoil and Plant Cover
Loss of Natural Drainage Patterns
Changes in Large River Flow Characteristic
Decrease Reserve Farmland
Decrease Groundwater Recharge
Stream/Land Erosion
Algal Blooms and Fish Kills
Endocrine Disruption in Aquatic Species and Feeders
Uptake of Contaminants in Biota/Food web
Riparian Habitat Loss

Table 3. Category 3: Water Contaminations Problems

First Storm Surge Toxic Materials
Impervious Topping Compounds
Acid Mine Drainage
Nitrates
Low-High pH
Oil/Grease
Persistent Organic Compounds
Combined/Sanitary Sewer Overflows
Increase in Pet Fecal Matter
Methylmercury
Arsenic, and Other Heavy Metal Contaminants
Organohalogen Compounds
Low Dissolved Oxygen Levels/High BOD
High Turbidity/Dissolved Solids

land erosion, riparian habitat loss, and loss of groundwater recharge potential diminish their ability to naturally mitigate Category 1 problems.

For instance, there is a direct relationship between a healthy ecosystem and its ability to help hold and purify water. So category 2 decrements exacerbate or cannot help decontaminate category 3 water contamination problems and cannot aid in the retention of water following storms or during periods of snowmelt. An ecosystem consists of the physical environment and all bacteria, fungi, plants and animals that inhabit it. An ecosystem recycles mass -so organic pollutants are degraded using the same mechanical processes and reduction-oxidation (redox) reactions that are used to degrade natural organic matter (Hemond and Fechner-Levy, 2000). These reactions, performed on organic compounds (natural and anthropogenic) over and over again by aerobic and anaerobic bacteria and fungi in association with worms, insects and other organisms, tend to produce carbon dioxide, water and mineral salts. This process is known as mineralization.

Sometimes biotransformation of organic pollutants is not complete, termed partial biodegradation. But, even when an organic pollutant is only partially biodegraded the resulting chemical product is usually less toxic than the parent compound (There are notable exceptions to this general sequence. In anaerobic soil conditions trichloroethene {TCE} is biotransformation first to a series of three dichloroethene isomers, bacteria can further reduce these isomers to vinyl chloride, a known human carcinogen. Additionally, metallic mercury is biotransformed by the methylating action of bacteria into the fetal and neuro-toxin methyl mercury). Tree and plant cover take-up and hold rainfall and intact soils are engines of necessary biological activity. An undisturbed vadose or unsaturated zone assures slow transport and bacterial cleaning, via biofilms on particles in the porous subsurface media, of infiltrated water into lakes and streams and reliable recharge of groundwater aquifers.

So disturbance and loss of ecosystem services in category 2 has a profound effect on the types, concentrations and movements of introduced category 3 contaminants and thus the effects of primary Category 1 problems as well as the over-riding hydrological cycle. Historically, the agricultural methods used in the first human civilizations in the Fertile Crescent combined with deforestation, and river canalization led to changes in the climate in the area and soil salination. The result is that the Fertile Crescent is not fertile anymore and there has been a steady march north and westward of the center of political power to areas that are have abundant rainfall. In Europe and North America widespread development and loss of ecosystem services can create serious cross-border and intra-governmental disputes. The top soil reused to cover the disturbed sub-soils in residential and commercial land uses in developed countries are often not pre-development depth and are planted with a monoculture of grass; trees and shrubs. These plantings generally are of insufficient height and root structure to offer much help in water retention; non-native species are also introduced. Soils no longer contain as much biological capacity to degrade introduced pollutants and the transport times of contaminants to streams, lakes and main stem rivers are shortened due to stormwater runoff, inadequate surface infiltration, slope changes, and loss of riparian habitat. It takes years if not generations to build the type of biological activity that the surrounding ecosystem once provided.

Secondary Water Management Outcomes

Primary, category 1 issues act through complex interactions with preexisting ecological conditions and/or new category 2 conditions that they cause, combined with new or existing category 3 contamination issues to cause category 4 secondary water management problems. An important secondary water management problem associated with the example of development in critical watersheds and sprawl especially in developed countries is increased stormwater and/or snowmelt runoff. Modern stormwater engineering practices are designed to move water away from new developments as swiftly as possible into receiving streams and rivers. While this protects those in the develop-

Table 4. Category 4: Secondary Water Management Outcomes

Category 4: Secondary Water Management Outcomes
Human Pathogens in Surface Water
Human Pathogens in Groundwater
Increase Potential-Mine Blowouts
Increase Sediments in Surface Water
Decreased Production Clean Surface/Groundwater
Habitat Loss/Fracture
Increased Storm-water/Snowmelt/Runoff
Increased Contaminant Loads Surface/Groundwater
Consumption of Contaminated Fish
Flooding
↓ Confined /Unconfined Aquifers/Storage Ability
Human Exposure: – Carcinogens – Toxic Substances – Endocrine-Active Substances

ment and their immediate neighbors it causes flooding at critical discharge points downstream (French-University of Pittsburgh Institute of Politics, 2006). Downstream the incidence and severity of floods is increasing very often in older urban and suburban communities. Development in critical watersheds and sprawl decreases the production of clean surface and groundwater, adding to the burden of contaminants in surface and groundwater and ultimately exposure of humans and aquatic and terrestrial species to toxic and hazardous substances via ingestion and skin absorption (fish are exposed through their gills). There is an increase in water sediments as a result of erosion. Human pathogens enter receiving waters and main stem rivers from combined and sanitary sewer and wastewater treatment plant overflows as well as from entrained pet fecal matter. Groundwater can also become contaminated by pathogens resulting in exposure to well water users, well water is not regularly tested or treated for pathogens.

Tertiary Water Management Outcomes

If left unchecked these secondary water management outcomes can singularly or in combinations cause category 5 tertiary water management outcomes. Flooding can lead directly to the loss of human life, personal injury and property damage. The social and economic underpinnings of communities are disrupted and psychological disturbances including posttraumatic stress disorder have been reported. Increasing dollars are required to be put into downstream stormwater management and flood insurance programs; costs not borne by the original land developers and new inhabitants but by the general public. Increased sediments in water along with contaminants and pathogens add to the cost of water purification, again usually paid for by the general public. Even in areas of the world that have a surplus of water, these water quality problems can have an impact on available water quantity and ultimately sustainable economic growth.

Throughout the world anglers, alone or in unorganized or organized groups, use streams, lakes and rivers and estuaries for subsistence style fishing as well as recreational purposes. Bio-concentration of contaminants in fish can cause exposure of subsistence and commercial fish eaters to heavy metals, environmentally persistent organic compounds (polychlorinated biphenyls [PCB's], DDT and dioxins for example) and xeno-estrogens via ingestion (Miyamoto J., and J. Burger. 2003). This raises human risk for the development of cancers as well as other environmental diseases. Destruction or degradation of water resources also greatly diminishes their recreation potential. It is unwise to compromise outlets for human physical activity as evidenced by the epidemic of chronic obesity seen in the USA. Habitat loss and fracture from developments in critical watershed can not only lead to animal population declines and loss of native aquatic and terrestrial plants but to aesthetically unpleasant landscapes. Aesthetically pleasing landscapes impart a feeling of well being in humans and their loss could result in a buildup off stress. There is a strong correlation between the occurrence of both high monthly precipitation and wet weather events and disease outbreaks (Rose et al., 2000) (Curriero et al., 2001). Surface water outbreaks occurred most often in the month following the wet weather event and groundwater outbreaks were associated with a 2-month lag period between the precipitation event and waterborne disease outbreaks. Studies at and downstream from combined sewer overflows (CSO's-they combine sanitary sewage and stormwater drains) outfalls in main stem rivers have shown elevated levels of Cryptosporidium and even higher levels of Giardia (States et al., 1997) (Gibson et al., 1998). This poses a risk to drinking water and those coming in contact with the water while fishing or other recreational activities. Fisherman have reported gastro-intestinal disturbances following water contact subsequent to seeing CSO gates open during wet weather events (Volz, 2006).

Pathogens in water remain a major cause of water-borne illness throughout the world. In the USA-Cryptosporidium parvum, a human and animal intracellular parasite has been a poster child for waterborne illness since the late 1980's. Its oocysts are particularly resistant to common water purification treatments (Robertson et al., 1992). A massive outbreak in Milwaukee of Cryptosporidium infection transmitted through the public water supply affected approximately 400,000 people with mild, moderate and severe watery diarrhea in 1993 (Mackenzie et al., 1994), deaths among the imunocompromised were reported (Hoxie, 1997) and the mortality rate among infected, imunocompromised individuals was estimated to be over 50% (Rose, 1997). The outbreak resulted in an estimated total cost of over US $93 million including direct medical costs and productivity losses (Corso, 2003). This outbreak was associated with high water runoff from snowmelt and precipitation, high water turbidity at water intakes, and a failure of the water filtration system.

There is a strong correlation between the occurrence of both high monthly precipitation and wet weather events and disease outbreaks (Rose et al., 2000) (Curriero et al., 2001). Surface water outbreaks occurred most often in the month following the wet weather event and groundwater outbreaks were associated with a 2-month lag period between the precipitation event and waterborne disease outbreaks. Studies at and downstream from combined sewer overflows (CSO's-they combine sanitary sewage and stormwater drains) outfalls in main stem rivers have shown elevated levels of Cryptosporidium and even higher levels of Giardia (States et al., 1997) (Gibson et al., 1998). This poses a risk to drinking water and those coming in contact with the water while fishing or other recreational activities. Fisherman have reported gastro-intestinal

Table 5. Category 5: Tertiary Public Health, Social, Emotional and Economic Outcomes

Property-Flood Damage
Flood-Loss of Human Life
↑ Storm -water Management Costs
↑ Costs Water Purification
↓ Human Aesthetic Value
↓ Recreation Potential
↓ Economic Growth
Loss of Aquatic/Terrestrial Species
↑ Costs-Flood Protection/Insurance
↑ risk Cancer/Humans
↑ Risk Other Environmental Disease
↓ In Water Quantity
↑ In waterborne Pathogen Disease

disturbances following water contact subsequent to seeing CSO gates open during wet weather events (Volz, 2006).

Breaking the Chain of Causation – Focusing on Restoration and Protection of Ecosystem Services; An Important NATO Mission

The entire argument of this paper is to engage in primary preventative measures to combat primary water management problems. Often times though prevention of a primary problem such as use of marginal cropland/upslope lands for agriculture or mass movement of rural populations to cities with development in critical watersheds cannot be addressed directly without placing a severe political or economic strain on a country. Protection or restoration of ecosystem services can help mitigate the effects of long-term environmental degradation while programs for sustainable growth are developed. NATO has many of the engineering, public health, medical and logistical services necessary to assist countries and regions with ecosystem rehabilitation service. Large scale reforestation, wetlands construction, appropriate dam removal, native species plantings, and land and water-course erosion control projects would pay substantial dividend for prevention of secondary and tertiary water problems. If NATO would use its resources in this fashion in a major way it would go far to fulfilling its mission of peace maintenance. Lastly, regional, inter country holistic watershed management is necessary to break the chain of causation at the primary prevention level stage to stop secondary water outcomes and prevent further tertiary public health, economic and social-behavioral consequences. Since flooding and other tertiary problems do not follow political boundaries but watershed boundaries, solutions to these primary problems will take a combined effort of groups of governments.

CDV is Principal Investigator for the Pittsburgh Fish Consumption Study supported by a grant from the DSF Charitable Trust through the UPCI-CEO. His water policy work is supported by a grant from the Heinz Endowment through the CHEC and by the University of Pittsburgh, Institute of Politics. CDV is a Co-Investigator, University of Pittsburgh Academic Center of Excellence (UPACE) for Environmental Public Health Tracking (EPHT). He was the Director of the 2004 Amchitka Expedition to monitor radionuclides in the marine environment as a result of nuclear detonations on the island in the late 1960's and early 1970's. Many thanks to Nancy Sussman, PhD and Yan Liu, BS Env. Eng. for their review of this manuscript. Thanks also to Carol Larach, MPH for her perseverance in making this project possible.

References

[1] Corso PS, Kramer MH, Blair KA, Addis DG, Davis JP and Haddix JC. 2003. Cost of illness in the 1993 waterborne Cryptosporidium outbreak, Milwaukee, Wisconsin. Emerging Infectious Disease; 9 (4): 426–31.

[2] Curriero, F., J. Patz, J. Rose, and S. Lele. 2001. The association between extreme precipitation and waterborne disease outbreaks in the United States, 1948–1994. American Journal of Public Health 91(8): 1194–1199.

[3] French DR 2006. The rising floods: why Southwestern Pennsylvania's flood problems are worsening in University of Pittsburgh, Environment Committee, Water Management Framing Paper.

[4] Gibson, C., K. Stadterman, S. States, and J. Sykora. 1998. Combined sewer overflows: A source of *Cryptosporidium* and *Giardia*. Water Science and Technology 38(12): 67–72.

[5] Hemond HF and Fechner-Levy EJ. 2000. *The chemical fate and transport in the environment, 2nd edition.* San Diego, CA: Academic Press.

[6] Hoxie NJ, Davis JP, Vergerint JM, Nashold RD and Blair KA. 1997 Cryptosporidiosis-associated mortality following a massive waterborne outbreak in Milwaukee, Wisconsin. American Journal of Public Health.Dec; 87(12): 2032–5.

[7] Mackenzie WR, Hoxie NJ, Procter ME, Gradus MS, Blasir KA, Peterson DE, Kazmerski JJ, Addis DG, Fox KR, Rose JR et al., 1994. A massive outbreak in Milwaukee of cryptosporidium infection transmitted through the public water supply. New England Journal of Medicine Jul 21; 331(3): 161–7.

[8] Miyamoto J., and J. Burger. 2003. Implications of endocrine active substances for humans and wildlife: executive summary. Pure and Applied Chemistry .75, nos. 11–12, xv–xxiii.

[9] Naiman, R. J., Magnuson, J. J., McKnight, D. M., and Stanfield, J. A. 1995. *The freshwater imperative. A research agenda.* Washington, DC: Island Press.

[10] Robertson LJ, Campbell, AT and Smith HV, 1992. Survival of Cryptosporidium parvum oocysts under various environmental pressures. Applied Environmental Microbiology; 58(11): 3494–500.

[11] Ritter L, Solomon K, Sibley P, Hall K, Keen P, Mattu G, and Linton B. 2002. Sources, pathways and relative risks of contaminants in surface and groundwater: a perspective prepared for the Walkerton inquiry. Journal of Toxicology and Environmental Health. Part A; Jan 11; 65(1): 1–142.

[12] NRC (National Research Council). 1999. New Strategies for America's Watersheds. Washington, DC.

[13] Rose J.B. 1997. Environmental ecology of Cryptosporidium and public health implications. Annual Review of Public Health; 18: 135–61.

[14] Rose, J., S. Daeschner, D. Easterling, F. Curriero, S. Lele, and J.A. Patz. 2000. Climate and waterborne disease outbreaks. Journal of the American Water Works Association 92(9): 77–87.

[15] States, S., K. Stadterman, L. Ammon, P. Vogel, J. Baldizar, D. Wright, L. Conley, and J. Sykora. 1997. Protozoa in river water: Sources, occurrence, and treatment. Journal of the American Water Works Association 89(9): 74–83.

[16] Volz, C. D. and Christen, C., 2006. Unpublished focus group interviews with "meat" and sport fishers in the Three Rivers Area of Pittsburgh, The Pittsburgh Fish Contaminant Study.

Strengthening National Public Health Preparedness and Response to Chemical, Biological and Radiological Agent Threats. Edited by C.E. Cummings and E. Stikova.
IOS Press, 2007.

Social Marketing as a Potentially Valuable Tool for Preparedness

Peter D. RUMM [a] and Curtis E. CUMMINGS [b]
[a] *Center for Devices and Radiological Health, US Food and Drug Administration, 9200 Corporate Drive, Rockville, MD 20850 (formerly, Drexel University)*
[b] *Drexel University School of Public Health, MS 660, Philadelphia, PA 19102-1192, USA*

Abstract. A major component of the North Atlantic Treaty Organization (NATO) Advanced Study Institute (ASI), *Strengthening National Public Health Preparedness and Response to Chemical, Biological, and Radiological Agent Threats,* in June, 2006, focused on preparing for and implementing communication strategies for emergency preparedness. During the ASI, speakers stressed the need for better communication in planning, training and evaluating exercises and actual responses to both man-made and natural disasters. To date, communication strategies have mostly emphasized risk communication. For example, during the early years of the bioterrorism funding (2002 to 2004), the U.S. Centers for Disease Control and Prevention (CDC) funded preparedness programs according to Focus Area categories, one of which, Focus Area F, was devoted to risk communication [1], and this area continues to be crucial to public health communication. At the ASI, an additional topic introduced was the emerging field of social marketing as an effective strategy for preparedness, especially in the United States.

Social Marketing and Its Usefulness in Preparedness

Simply, social marketing is the use of commercial advertising techniques to promote effective health messages. [2] Andreasen's (1995) frequently-quoted definition of social marketing is the "application of commercial marketing technologies to the analysis, planning, execution, and evaluation of programs designed to influence the voluntary behavior of target audiences in order to improve their personal welfare and that of their society." [3] In a public health or preparedness setting, social marketing aims to achieve social goals through behaviors that improve health outcomes, as does all health-related communication.

In 1972, Kotler and Zaltman first proposed that standard marketing principles could be used to promote ideas, attitudes and behaviors that benefit target audiences and society. [4] In public health, a large body of literature and experience has proven them right, in areas that range from control of tobacco use to reduction of workforce hazards.

To effectively "sell" healthy behavior, social marketing should start with audience research that divides the target audience into groups that have common risk behaviors, motivations, and information channel preferences [5]. Certain audience segments can then be reached with a mix of intervention strategies that are guided by the "4 P's" of marketing [5]:

- **Product:** what the consumer is asked to "buy" (often a behavior)
- **Price:** the cost or something the consumer must give up/ do to obtain the product
- **Place:** how and where the product reaches the consumer
- **Promotion:** how information about the product is disseminated

Additional information and links to societies working in the field of social marketing can be found at the CDC web site, at a web page devoted to this growing area on how to employ these techniques to plan a campaign. It is an area that requires experience and training, and the CDC and others offer courses in this area. For this purpose, the CDC recently founded a new center called the National Center for Health Marketing. [5]

At any level of government (local, regional or national), it is critical that all health communication materials and campaigns be carefully planned, and then get "vetted" in focus groups or pilot studies to evaluate their content – a lack of which is often the reason that some campaigns have failed in the past. [6] There also should also be frequent, even continual evaluation of marketing efforts once they are implemented.

The following are recent examples of how social marketing has been used in the U.S.:

1. In 2005, the Center for Public Health Readiness and Communication (CPHRC) at the Drexel University School of Public Health in the U.S. worked with an African American community in the Eastwick neighborhood of Philadelphia, PA. The community's advocacy coalition, the Eastwick Project Area Committee, expressed concern about a potential release of chemicals from a large adjacent petroleum refinery. The CPHRC held focus groups and developed and began the dissemination of messages to broadcast during either a natural or man-made disaster, including a chemical emergency at the refinery. [7,8]
2. In 2004, the State of Rhode Island, in the U.S., launched the first of its kind, state-wide effort to distribute preparedness materials to its entire population. The state used CDC Focus Area F funding, spending about $200,000 for the entire project. Using focus groups and in-depth interviews by communication specialists, the state created information sheets in an acceptable mailed format that were sent to all 420,000 households in the state. [9]
3. Montgomery County, Maryland, U.S., received targeted funding from the National Association of County and City Health Officials (NACCHO) to develop culturally sensitive preparedness-related educational materials. Working with community members, the county prepared an emergency preparedness check list for local citizens termed "Plan 9," available both on-line and in print, in many languages. This work has been seen as an effective model for use by other U.S. cities and counties. [10]

As health departments, academic institutions and commercial entities move into the field of social marketing for preparedness, they will gain experience and learn lessons. This experience should help evolve effective strategies and provide a body of knowledge on how best to conduct and effect change through messages that are tailored to each audience. Well-written messages have the potential to save lives, especially in the preparatory and prevention stages of natural and man-made disasters. Finally, the

Supercourse is one expanding global medium that can be used to increase knowledge regarding social marketing for preparedness. [11] It can be promoted as a cost-effective conduit by which social marketing can be introduced to the public health community world-wide.

References

[1] Emergency Preparedness and Response. U.S. Centers for Disease Control and Prevention (CDC), Department of Health and Human Services. http://www.bt.cdc.gov/erc/.

[2] T Evans, President, Educational Messaging Services, Inc., Ventura, California, USA. Personal communication, 12 November 2006.

[3] Andreasen, A. 1995. Marketing Social Marketing in the Social Change Market Place, Journal of Public Policy and Marketing, Vol. 21, No. 1, 3–13.

[4] Kolter P and Zaltman G. 1972. *Journal of Marketing*, Vol. 36, No. 4, 60–61.

[5] U.S. Centers for Disease Control and Prevention, Department of Health and Human Services. http://www.cdc.gov/communication/practice/socialmarketing.htm.

[6] Rumm P. Social Marketing the Science and its Evaluation, Grand Rounds, Drexel University School of Public Health, 21 February 2005.

[7] Rumm P, Evans T, Sanderson A, Sinibaldi J, Vaughn N, Villanueva A and Cummings C. A draft community (Eastwick, PA) based social marketing campaign for preparedness. Presented as part of the Eastwick Community Emergency Preparedness Project, Philadelphia, Pennsylvania, USA, 11 August 2005.

[8] Rumm P, Sczersputowski J. A model social marketing campaign for terrorism preparedness. American Public Health Association Annual Meeting, Philadelphia, Pennsylvania, USA. 12 December 2005. Poster session.

[9] Rhode Island Department of Health. Personal communication.

[10] Montgomery County Department of Health and Human Services Section on Emergency Preparedness, cited at: http://www.montgomerycountymd.gov/mcgtmpl.asp?url=/content/PIO/news/preparedness.asp.

[11] Linkov F, LaPorte R, Sauer F, & Shubnikov E. Public Health Preparedness: I-Prevention and Global Health Network Supercourse. NATO Advanced Study Institute: Strengthening National Public Health Preparedness and Response for Chemical, Biological and Radiological Agent Threats. June 19–29, 2006.

Strengthening National Public Health Preparedness and Response to Chemical, Biological and Radiological Agent Threats. Edited by C.E. Cummings and E. Stikova.
IOS Press, 2007.

Including Diverse Populations with Unique Needs in Emergency Planning

Carol S. LARACH MSEd, MPH, Curtis E. CUMMINGS MD, MPH
and Marcia POLANSKY ScD, MSW
Drexel University School of Public Health MS 660
Philadelphia, PA 19102-1192, USA

Abstract. Public health approaches to emergency preparedness and response have evolved to become more inclusive of special needs and vulnerable populations. Despites these advances, more effective ways of addressing emergency preparedness for these populations are needed. Ways of defining and identifying special needs and vulnerable populations which require special attention during a disaster will be discussed. Strategies for including them in the emergency preparedness planning process, effective approaches to communication and training of public health response personnel for working with special needs and vulnerable populations will also be addressed.

Keywords. Emergency preparedness; diverse populations; special needs; vulnerable populations; hazards

1. Introduction

Public health systems around the world have begun to address emergency preparedness for special needs and vulnerable populations. However, gaps in emergency preparedness for these populations remain [1]. An effective approach for these groups requires the full involvement of all components of the public health system as well as the identification of all groups who require special consideration so that resources can be appropriately allocated [1]. Effectively addressing these populations' needs during disasters requires that disaster preparedness and response organizations work effectively with these populations and their advocacy groups.

Overlooking populations with special needs in emergency preparedness has led to catastrophic consequences in the past. Recent disasters, such as the terrorist attacks of September 11, 2001, the 2004 Indian Ocean tsunami, and the 2005 Hurricane Katrina, revealed the vulnerability of special populations. For example, during the World Trade Center Disaster (2001), many individuals with disabilities and medical conditions, such as those requiring wheelchairs, walkers, canes, and oxygen, were not able to make their way out of the towers and died [2]. Similarly, in New Orleans, Louisiana, 71% of the fatalities from Hurricane Katrina were elderly people, age 60 or older, who resided in nursing home facilities [3]. Adults ages 50 or older were also over represented among the victims of the 2004 tsunami in Sri Lanka, comprising 15.3% of the fatalities [4]. Sadly, children were the most vulnerable of all and suffered the most fatalities from the tsunami (31.8% of all fatalities).

The aim of this report is twofold: (1) to discuss the advantages and disadvantages of the various definitions of special needs and vulnerable populations used by disaster preparedness and response organizations and (2) to address effective strategies for emergency preparedness for special needs and vulnerable populations that would insure emergency preparedness for all people.

2. Terminology of "Special" and "Vulnerable" Populations

Disaster preparedness and response organizations as well as researchers in public health preparedness have put forth a range of views as to what constitutes special needs or vulnerable populations. For example, the U.S. Federal Emergency Management Agency (FEMA) uses the term "special needs" to describe "individuals in the community with physical, mental, or medical care needs who may require assistance before, during, and/or after a disaster or emergency after exhausting their usual resources and support network."[5] FEMA includes certain individuals with sensory disabilities, such as reduction or loss of sight or hearing; with mobility disabilities, such as use of a cane or wheelchair; with mental disabilities, such as persons who are mentally ill or developmentally disabled or those with traumatic brain injury or learning disabilities; and with other medical conditions, such as renal dialysis, diabetes, lung diseases requiring oxygen and epilepsy. The Research and Training Center (RTC) on Independent Living at the University of Kansas, describes those with special needs as "people with disabilities (including cognitive, physical, and sensory); age spectrum (pediatric to geriatric); and those requiring medical assistance (home care, assisted living)" [6]. RTC's definition has an advantage over the FEMA definition in that it explicitly identifies certain age groups that are at high risk due to the prevalence of people with special health issues among them. However, advocacy organizations of special populations have been critical of the definitions used by these and other public agencies. One concern of the advocacy groups is that the current definitions and designations of special needs and vulnerability to disasters do not adequately encompass the entire disability community [7]. Another concern is that culture, ethnicity, language, and socioeconomic status may not be considered as leading to inadequate protection from catastrophes for some populations.

There also has been research on the effect of hazards and disasters from a global perspective; specifically vulnerability caused by political, social and economic factors [8]. Within this context, populations at risk are defined as "groups historically disadvantaged by socio-economic status, patterns of discrimination and/or exclusion, lack of political representation and/or cultural distancing" [9]. Included in this definition are ethnic minorities (by language); immigrants and migrants; low-income households; residents of group living facilities; women-headed households; renters; physically or mentally disabled; the homeless; and tourists and transients [10]. While more inclusive, this definition may overlook some groups and by specifically naming certain groups, overemphasizes them. Sorensen suggests that there are vulnerable individuals who are not disabled physically or mentally but are more vulnerable due to their marginalized position in society. They are "less visible to agencies or officials in emergencies and less likely to self-identify before a hazardous event" [8]. The homeless, for example, tend to be isolated from service systems and may only be known by law enforcement or specific organizations [8].

As there is not yet a consensus as to specifically which groups should be included in the special needs/vulnerable category, the question arises as to whether there is an alternative approach to defining special needs/vulnerable populations. Sorensen suggests characterizing vulnerable groups as those groups "whose needs are not fully addressed by traditional service providers" [8]. This approach may allow more flexibility in responding to specific disasters. Certain group(s) might be vulnerable to a certain type of disaster but not to other disasters or the vulnerability of a certain group to a disaster might be unexpected. With Sorensen's definition, a group could be immediately designated as vulnerable and become eligible for assistance. In the U.S., the organization PrepareNow.org in the state of California has extended its definition of vulnerable populations to include "people who feel they cannot comfortably or safely access and use the standard resources offered in disaster preparedness, relief and recovery" [8]. This extended definition is important since some individuals have special needs in disasters which are not obvious or which could be anticipated. Consequently, these individuals might not receive special services they require if they were not encouraged to self-identify. Kailes and Enders espouse a function based framework for emergency management and planning based on essential functional needs [11]. They include medical needs, communication needs, supervision needs, maintaining functional independence, and transportation as functional needs .The advantage of addressing functional limitations is that it identifies individuals who have a disability, as well as a larger range of individuals who do not identify as having a disability, but do have some level of limitation. For example, an individual who has a broken bone which although temporary would interfere with their functioning during a disaster.

3. Strategies

A review of the disaster and hazards literature suggests that the following strategies may enable planners to ensure that these diverse vulnerable and special needs populations are fully included in emergency preparedness.

3.1. Identify the Special Needs or Vulnerable Populations at the Local Level

Phillips and Morrow suggest that local emergency preparedness systems identify these special-needs or vulnerable populations within their communities or jurisdictions [9]. In the U.S., data sources to identify these populations include census data, special censes/registries, county health department and social service agencies, advocacy organizations, and faith-based organizations [8]. For example, a county health department generally has information regarding people with disabilities, institutionalized groups, and the elderly. Faith-based organizations often have information regarding migrants and the culturally isolated. Census data can be used to identify census tracts which have a high proportion of residents below the poverty line and those with overcrowding and inadequate housing. This combination of adverse conditions makes these communities vulnerable to a disaster. The United Nations and the International Marketing Data and Statistics have data on these groups globally [12].

Having maps that display these populations available to emergency responders is an invaluable aid for their field work. Options to develop useful maps range from simple, low-tech maps to locate high-risk groups, to high-tech, Geographic Information

Systems (GIS) based tools that integrate several types of data to more specifically locate those with multiple vulnerabilities [10].

3.2. Engage Special Needs and Vulnerable Populations and Their Advocates in the Planning Process

To better meet the preparedness needs of special and vulnerable populations, emergency management organizations need to solicit and incorporate input from these populations and their advocates during the planning processes. The accumulated self-knowledge and expertise of these groups in dealing with their special needs is invaluable for emergency management organizations to build upon. In addition, their inclusion in the planning process increases the trust in and the credibility of the preparedness plans among those who will be the recipients. Specifically, the input from these populations can be used to "tailor the planning initiatives and drills", as well as "how to best reach out to these communities to increase awareness on emergency preparedness. [8,13]. Creating a special needs task force within emergency management organizations can be very effective in maximizing input from the special needs and vulnerable populations. These steps help make special needs and vulnerable populations "full partners" in emergency planning and preparedness [13].

3.3. Training of Emergency Responders

Emergency responders require specialized training to better understand the unique needs of special needs and vulnerable populations and on how to assist them during emergencies. As was evident during recent disasters, emergency responders are generally ill prepared to assist and serve these populations. For example, after the Northridge, California, earthquake in 1994, response personnel misidentified disabilities as acute medical conditions and inappropriately referred individuals with disabilities to medical facilities [2]. During the evacuation phase of Hurricane Katrina, individuals with psychiatric disabilities were mistreated and inappropriately institutionalized as a result of first responders not having training on disability issues [14].

Required skills to address these populations effectively include how to decrease anxiety levels of emotional people, recognizing cognitive impairments, and evacuate and transport, lift or carry people with disabilities and assistance animals [8,15]. Cameron suggests that when providing training, it is extremely valuable to use members of these populations as trainers, as they possess "concrete, practical knowledge" [15].

Also, preparedness training should be extended to the special needs and vulnerable populations. They have considerable personal experience with coping with a disability and/or functional limitations. They also have ongoing relationships with others with disabilities, so they can, in turn, train their own "communities" as well as disseminate emergency preparedness information.

4. Exercise the Plans – Table Top, Drills, and Full-Scale

To examine whether or not the concerns of special needs and vulnerable populations have been adequately addressed and integrated, Davis and Mincin suggest regularly testing the disaster plans via table top, functional, and full-scale or real-life/real-time exercises [6]. Members of these populations, as well as, experts, advocacy groups, and leaders of these groups should be involved in these preparedness exercises as both observers and participants. By "testing the systems," gaps can be identified and existing plans can be modified [6]. In 1997, the New York Office of Emergency Management incorporated disability issues in an urban terrorism drill, Interagency Chemical Exercise, to ensure that first responders were presented with a realistic situation and that emergency response organizations improve their systems [6]. People with disabilities participated in the exercise. These "victims" provided feedback on how they were treated and what the overall experience was from a disability perspective, and identified strengths and weaknesses of the response system.

5. Communication

Communication is another key issue to address. A wider variety of approaches are needed to adequately address the diversity among these populations [9]. Emergency preparedness information needs to be available in alternate formats, including audiotape, electronic, and written material in large type, Braille, and languages other than English [6, 9]. Similarly, Public Service Announcements, forecasting and warnings should be in Captioned (open and closed) for people who are deaf or hard of hearing and in various languages for the non-English speakers [6,12]. For non-English speakers, "other-language" radio and TV channels should be used to broadcasting announcements, forecasts, and warnings [12]. In the U.S., a disaster where language interfered with response and recovery was the 1987 Saragosa, Texas tornado [9,12]. Many Spanish-speakers perished as a result of the Spanish-speaking language television stations not broadcasting the local warnings. The local station did broadcast, but incorrectly translated the word "warning" from English to Spanish [9]. Locating translators for every language, including sign language, in the community is essential. Thus, to effectively communicate with these populations, it is of paramount importance that the planners understand how these groups receive, interpret and respond to information; an area that requires further research [9].

Conclusion

Special needs and vulnerable populations include many diverse groups, each with unique attributes and requiring a preparedness approach specifically tailored to address their special needs and vulnerabilities. To ensure the safety and survival of all people during times of disaster, emergency preparedness organizations and planners must broaden their definition of special and vulnerable populations. Agencies involved in public health preparedness need to learn from each other or from special needs and vulnerable populations. We have described some of the elements which need to be incorporated into preparedness program to adequately address preparedness for special needs and vulnerable populations.

References

[1] Noji, E.K. (2005). Disaster: Introduction and State of the Art. *Epidemiologic Reviews,* Vol. 27, No. 1 pp. 3–8.

[2] National Council on Disability (2005). *Saving lives: including people with disabilities in emergency planning.*
[3] Hurricane Katrina: Lesson Learned. http://www.whitehouse.gov/reports/katrina-lesson-learned/chapter1.html.
[4] Nishikiori, N., Abe, T., Costa, D.G., Dharmaratne, S.D., Kunii, O., & Moji, K. (2006). Who died as a result of the tsunami? – Risk factor of mortality among internally displaced person in Sri Lanka: a retrospective cohort analysis.
[5] Emergency Management Institute, EMI-6197 Emergency Planning and Special Needs Populations – Instructors Guide (2003). http://training.fema.gov/EMIWed/pub/downloads.asp.
[6] Davis, E. & Mincin, J. (2005). *Nobody left behind: Incorporating special needs populations into emergency planning and exercise.* www.nobodyleftbehind2.org.
[7] Kailes, J.I. (2005) in Davis, E. & Mincin, J. (2005). *Nobody left behind: Incorporating special needs populations into emergency planning and exercise.* www.nobodyleftbehind2.org.
[8] Sorensen, B.V. (2006). *Populations with special needs.* Oak Ridge National Laboratory (ORNL/TM-2006/559).
[9] Phillips, B.D. & Morrow, B.H. (2005). *Social science research needs: a focus on vulnerable populations, forecasting and warning.* Paper presented at the Hurricane Forecast Socioeconomic Workshop, Pomona, CA.
[10] Morrow, B.H. (1999). Identifying and Mapping Community Vulnerability. *Disasters,* Vol. 23, No. 1, pp. 1–18.
[11] Kailes, J.I. & Ender, A. (in press). Moving Beyond "Special Needs": A Function Based Framework for Emergency Management and Planning.
[12] Enarson, E. (2004). *Social Vulnerability Course,* FEMA Higher Education Project. http://training.fema.gov/emiweb/edu/completeCourses.asp.
[13] National Organization on Disability (n.d.). *Report on special needs assessment for Katrina evacuees (SNAKE) project.* www.nod.org.
[14] National Council on Disability (2006). *The needs of people with psychiatric disabilities during and after hurricanes Katrina and Rita: position paper recommendations.* www.ncd.gov.
[15] Cameron, C.T. (n.d.). *Emergency preparedness for people with disabilities and other special needs; another look after Katrina.* Disability Preparedness Center.

Strengthening National Public Health Preparedness and Response to Chemical, Biological and Radiological Agent Threats. Edited by C.E. Cummings and E. Stikova.
IOS Press, 2007.

Author Index